Second-Hand Cultures and Economies of Reuse, Repair, Sharing, and Care

Petra Kuppinger
Editor

Second-Hand Cultures and Economies of Reuse, Repair, Sharing, and Care

palgrave
macmillan

Editor
Petra Kuppinger
Department of Sociology and Anthropology
Monmouth College
Monmouth, IL, USA

ISBN 978-3-031-99874-4 ISBN 978-3-031-99875-1 (eBook)
https://doi.org/10.1007/978-3-031-99875-1

Cover image credit : paula sierra/Getty Images.

This Palgrave Macmillan imprint is published by the registered company Springer Nature Switzerland AG.
The registered company address is: Gewerbestrasse 11, 6330 Cham, Switzerland

Contents

Notes on Contributors

Susan Andreatta is a Professor of Applied Cultural Anthropology at The University of North Carolina at Greensboro, USA. Her research and teaching interests coincide with an emphasis on social and environmental justice, food production, health, nutrition, and food security. She leads the sustainable campus garden initiative where she supervises 50 raised beds for faculty, students, and staff use. Since 2001, she has been the director for Project Green Leaf, a project that was established to assist farmers and the public in sustaining a local agriculture and food system.

Brieanne Berry is an Assistant Professor of Environment and Sustainability at Ursinus College. Trained as an environmental and economic anthropologist, Dr Berry has conducted research on Maine's vibrant reuse economies, and the value of localized thrift, resale, and redistributive economies in rural communities. She also has research interests in food waste and circular food systems, and serves on the leadership team of the Montgomery County Food Policy Council. Her work broadly explores the potential to find value in our discards. She authored more than a dozen peer-reviewed publications. Her interdisciplinary, collaborative research has been supported by the National Science Foundation and the Environmental Research and Education Foundation.

James S. Bielo is Associate Professor of Religious Studies at Northwestern University. He is the author of *Materializing the Bible: Scripture, Sensation, Place* (2021); *Ark Encounter: The Making of a Creationist Theme Park* (2018); *Anthropology of Religion: The Basics* (2015); *Emerging Evangelicals: Faith, Modernity, and the Desire for Authenticity* (2011); and *Words upon the Word: An Ethnography of Evangelical Group Bible Study* (2009).

Sujan Chakraborty earned his BBA and MBA, majoring in Marketing, from the University of Chittagong in Bangladesh. He worked for the central bank of Bangladesh for 4 years. His primary job responsibility there was to conduct bank inspections. After moving to the United States, he earned an MS in Economics from the University of Maine. During his stay in Maine, he did research on evolutionary economics, circular economy, and published a few papers revolving around these topics. Now he is pursuing a PhD in Economics at Indiana University Indianapolis focusing on the economics of philanthropy, and health economics.

Subhadra Mitra Channa is a retired professor of anthropology from Delhi University. Her research focuses on gender, cosmology, inequality and marginalization, identity, and urban studies. She is the author of *Gender in South Asia* (2013); *The Inner and Outer Selves* (2013); and *The Dhobis of Delhi: An Ethnography from the Margins* (2024). She co-edited *Life as a Dalit* (ed. 2016); *Religious Pluralism in India* (2023); and *Colonial Anthropology: Technologies and Discourses of Dominance* (2024). She was the Senior Vice President of the International Union of Anthropological and Ethnographic Studies (IUAES) and is at present the Co-Editor of *Reviews in Anthropology*. Her awards include two Fulbright teaching fellowships, Charles Wallace and several visiting professorships, as well as the Distinguished Teacher Award from Delhi University.

Simon Coleman is Chancellor Jackman Professor at the Department for the Study of Religion, University of Toronto, and a Fellow of the Royal Society of Canada. He is a former editor of the *Journal of the Royal Anthropological Institute* and a current co-editor of *Religion and Society*. Simon has carried out fieldwork in Sweden, Nigeria, and the United Kingdom. Recent books include *Powers of Pilgrimage: Religion in a World of Movement* (2022) and *Ideologies and Infrastructures of Religious Urbanization in Africa: Remaking the City* (2023, co-edited with David Garbin and Gareth Millington).

Andrew Crawley is an Associate Professor of Economics at the University of Maine. His research is focused on regional economic modeling, and the labor market, including matching efficiency analysis as well as input-output assessments. He has published in the leading regional science journals, *Regional Studies*, *Urban Studies*, and *Regional Science Policy and Practice*. His work has received funding from the National Science Foundation, National Institute for Food and Agriculture, National Oceanic and Atmospheric Administration, and the Economic Development Association. He also leads the EDA University Center at UMaine; this federally funded group supports regional economic research throughout the state of Maine working with public and private stakeholders.

Andrew Esiebo is a photographer based in Lagos who started his career by chronicling the rapid development of urban Nigeria. He is a winner of the Visa

Pour La Création Prize of the Institut Français and of the Musée du Quai Branly Artistic Creation Prize. See his work at andrewesiebo.com.

David Garbin is a Senior Lecturer in Sociology at the University of Kent, United Kingdom. His research focuses on migration, diaspora, transnationalism, urban change, and urban religion. He has conducted extensive fieldwork in Bangladesh, Nigeria, the Democratic Republic of Congo, Europe, and North America. He was the Principal Investigator of the "Pneuma-City" project, exploring the impact of end-of-life tires in West African megacities and the informal work of roadside tire repairers in Lagos (Nigeria). His books include *Ideologies and Infrastructures of Religious Urbanisation in Africa: Remaking the City* (2022, with Simon Coleman and Gareth Millington), *Religion and the Global City* (2017, with Anna Strhan), and *Migration and the Global Landscape of Religion* (2023).

Gretchen Herrmann is a retired librarian and anthropologist at SUNY Cortland where she taught Anthropology and Women's Studies. She earned her PhD in Anthropology from Binghamton University, and she has published numerous articles on garage sales, community, affect, reuse, local currencies, and Christmas games.

Cindy Isenhour is Professor of Anthropology and Climate Change at the University of Maine. Her research is focused at the intersection of environmental governance and economic policy, with particular attention to the concept of the circular economy and efforts to design more sustainable systems of production-consumption-disposal. She is the author of more than 60 peer-reviewed publications and three co-edited books including *Sustainability in the Global City* (2015, with Melissa Checker and Gary McDonogh); *Power and Politics in Sustainable Consumption Research* (2019, with Lucie Middlemiss and Mari Martiskainen); and *Consumption, Status and Sustainability* (2021, with Paul Roscoe).

Simon Johansson is an anthropologist affiliated with Stockholm University specializing in urban transformation and social inequality. He currently works with building evaluation capacity of the Swedish Agency of Defense Analysis. His dissertation, *Comeback Detroit*, examined the socio-economic shifts in Detroit as affluent residents "return" to revitalize the city, often displacing long-time marginalized communities. Johansson's academic work critically engages with the circular economy, gentrification, and the socio-political dynamics of post-industrial urban spaces, offering insights into how economic and cultural narratives shape cities and communities.

Erika Kuever is an associate professor of Consumption, Culture, and Commerce in the Department of Business and Management at the University of Southern Denmark, and has a background in Anthropology and East Asian Studies. Her research centers primarily around the politics of consumption in contemporary

China, including consumer rights, food safety, and provisioning under the COVID-19 lockdown. Some of her additional research interests include economic anthropology, tourism, religion and marketing, and consumption for children.

Petra Kuppinger teaches anthropology at Monmouth College. She has conducted research on space and power in Cairo, Egypt, and space, culture, and Islam in Stuttgart, Germany. Currently, she is researching sustainable projects, lifestyles, and urban cultures in Stuttgart. Her publications include *Faithfully Urban: Pious Muslims in a German City* (2015); *Cities and Places* (2023). She edited *Emergent Spaces: Change and Innovation in Small Urban Spaces* (2021). Her work has been published in *City and Society*; *Culture and Religion*; *City and Community; Anthropological Quarterly; Social and Cultural Geography*; *Journal of Urban Affairs*; *Space and Culture; Built Environment*; *Geographical Review*; *GeoJournal.*

Nicole Nathan is Faculty Director of Bachelor of General Studies at Temple University where she strives to make education more accessible, inclusive, and equitable for adult learners through adaptive pedagogy and expansive curricular development. Her research in cultural anthropology focuses on postcolonialism, religion, and tourism. Through a multi-sited approach to the study of transnationalism, her work seeks to delineate processes of globalization at local levels and centers marginalized voices. Her work on evangelical mission encounters highlights some of the problematic aspects of secondhand donations, including the ways in which charitable giving and consumerism form a mutually reinforcing system.

Samwel Moses Ntapanta is a postdoctoral fellow at the Department of Anthropology at Aarhus University in Denmark. He is an ethnographer of contemporary urbanism along the western Indian Ocean coast, focusing in particular on coloniality, consumption, and discarding, debris of late capitalism, repairing and recycling economies, and money. He is the author of *Gathering Electronic Waste in Tanzania: Labor, Value, and Toxicity.*

Henrietta K. Ransdell is a research fellow at the University of Louisville in Louisville, Kentucky. She previously completed a Fulbright grant in Chachoengsao Thailand and is completing her master's in interdisciplinary studies—Sustainability in May 2024. Her research interests include environmental anthropology, community engagement, and education.

Angela D. Storey is an Associate Professor in the Department of Anthropology and Director of the Anne Braden Institute for Social Justice Research at the University of Louisville (KY, United States). With an emphasis on collaborative and engaged scholarship, her research explores the politics of urban environments and experiences of community-based activism. She currently coordinates long-term ethnographic projects in Cape Town, South Africa, and Louisville.

List of Figures

List of Tables

CHAPTER 1

Introduction: Second-Hand Cultures and Economies of Reuse, Repair, Sharing, and Care

Petra Kuppinger

Fueled by overconsumption, inequalities in the neoliberal global economy, poverty, and concerns for the environment, recent decades have witnessed a robust continuation, revival, and growth of second-hand practices, markets, and cultures, and larger economies of reuse, repair, sharing, and care. Across the world, there are expanding landscapes of thrift stores, used goods street markets, swapping events, yard sales, flea markets, and Internet platforms like eBay or Facebook marketplace. There are global circuits of used clothes and appliances. Used or discarded appliances are repaired and resold across the world. There are global cyber networks that connect experts who repair cell phones and other electronics. Repair Cafés extend the lifespan of electronics, appliances, and textiles. Share boxes or shelves dot urban sidewalks. There are for-free stores and cafés and food-saving and sharing projects and apps. Recycling, upcycling, repurposing,

P. Kuppinger (✉)
Department of Sociology and Anthropology, Monmouth College,
Monmouth, IL, USA
e-mail: petra@monm.edu

P. Kuppinger (ed.), *Second-Hand Cultures and Economies of Reuse, Repair, Sharing, and Care*,
https://doi.org/10.1007/978-3-031-99875-1_1

and related practices that prolong the lifespan of objects are parts of everyday lives across the planet. Many of these practices have been robust elements of ordinary lifeworlds and cultures all along. Time-honored practices like fixing, mending, and creative reuse continue to play prominent roles in the lives of individuals and communities who engage in frugal, mindful, and resourceful practices that have been passed down to them. They stretch their resources and material possessions as much as possible because they have few other choices. Numerous individuals and communities make a living working with waste, and repairing, recycling, and upcycling things. People retrieve items and materials from waste streams. Simultaneously, practices of reuse, repair, sharing, and care are (re)discovered by younger generations or people who had previously considered them irrelevant or even a waste of time, because they were privileged enough to do so. These readopted practices are increasingly reaching the middle of wealthy (consumer) societies. Some have turned into profitable business ideas. Whether as the continuation of long-established practices, prompted by frugality, necessity, economic exclusion, long-held values and beliefs, or as practices fostered by environmental concerns, activities like second-hand markets, repair, repurposing, mending, or informal tool-sharing have proliferated in recent years. Guided by different motivations and necessities, people the world over buy and sell used goods, fix and resell appliances, or upcycle clothes and other materials for reuse. Some make art from cast-off objects or in creative ways recirculate the discards of others. The economic, social, and cultural circumstances and motivations for reuse, repair, sharing, and care differ vastly. Activities take endless shapes and are global in scope and impact. All help in their own small ways to reduce consumption, counter hyper-consumerism, challenge economic exclusion, and most importantly limit humanity's ecological footprint.

This book introduces people, spaces, and practices in this global landscape of second-hand cultures and economies of reuse, repair, sharing, and care. Chapters explore how practitioners join, participate, and expand global economies of reuse, repair, sharing, and care. People engage in daily work and routines that have always been sustainable like repairing household items, the sale of used items, or the creative reuse and repurposing of goods. They are mindful, creative, and careful in their use of material goods. Frugality and the abhorrence of waste are core values that guide long-established practices of recycling and reuse. Many individuals and communities are forced into second-hand economies, because they are excluded from dominant consumer economies and must buy used goods. Recycling and repair work are ways to make a precarious living in a global capitalist economy that litters the planet with the discards of the privileged. The necessity to divert goods from landfills and incinerators often guides individuals, institutions, and municipalities. Their quests are rooted in the understanding that there is much life left in many tossed goods, and in ecological or political ideas. Some individuals seek the cultural "cool" of thrifted items. Others avoid consumerism and see minimalism as a chance to shed environmental burdens and individual costs of hyper-consumerism. They insist that used goods serve their needs just as well or even better than new ones. Others see resistance to consumerism as political protest and seek to vote with their money. Buying used items, they avoid the dominant capitalist market and join more sustainable and circular economies. Participation in second-hand economies becomes a political indicator, and thrifted items can signify environmental politics and consciousness. In the Global North, greed might have been "cool" in the 1980s, but green and sustainable practices are a new cool for those who are eager to learn practices from their elders or elsewhere to limit consumption of new goods, extend limited resources, decrease their carbon footprint, and live simpler and more mindful lives. Across the globe, economies of reuse are vital elements for those who remain committed to values of frugality or simply must make ends meet. Economies of reuse, repair, sharing, and care are global in scope, but their expressions vary in different locations. Motivations to participate in these complex economies differ from involuntary to habitual, experimental, political, or activist. Diverse local and global second-hand circuits are interwoven.

This volume offers case studies of global second-hand cultures and economies and theorizes their complex interconnections. Much has been said about activities and projects in their specific context, but less work has been done to bring these examples into conversation with each other across the globe and across dividing lines between wealthy and less wealthy countries and lines of economic privilege, marginalization, and exclusion. Second-hand economies produce connections via the exchange of used goods, even if these lines remain invisible and are rarely acknowledged, and only occasionally problematized. This book juxtaposes case studies, activities, and projects and addresses questions about second-hand cultures from the diversion of consumer goods from landfills via municipal sharing websites, to the teaching of practices like avoiding single-use plastic on a farmers' market, to discussions of the landscape of used tires in a megalopolis like Lagos. Authors address the problem of donated clothes being forced onto recipients who do not need or want them and are often offended by these items, their quality, and this overall mode of "giving." Authors address the material and spiritual quality of religious material items and how they are resold in physical markets or via the Internet. Authors ask crucial theoretical questions about the role and potential of small projects to initiate change and contribute to urban, national, and global political transformations.

Chapters explore activities in North America, Africa, Europe, Asia, and the Caribbean. Authors pay attention to the complexities of concrete social, cultural, and economic circumstances where activities unfold and the historical dynamics that shaped them. They describe processes of reuse, repair, sharing, and care and explore practitioners' motivations. Based on ethnographic research, contributors describe details of practices and interactions in spaces and markets of reuse, repair, sharing, and care. They talk about the knowledge and skills involved in these practices. What values guide people's actions? What do social spaces look like where economies of reuse, repair, sharing, and care unfold? How do socialities in these spaces differ from those of the dominant capitalist economy? Authors inquire about differences between garage sales and brick and mortar thrift stores and their respective socialites. They discuss practices of reuse and recycling at a farmers' market and how these practices affect interactions and (learning) experiences at the market. Can farmers' markets encourage green practices beyond the sale of vegetables, and attempt to create more engaged, aware, and sustainable communities? What motivates women of different classes in India in their practices of reuse, repair, or recycling of saris? How do they rework or repurpose these fabrics? Contributors examine the intersection of the sacred and profane in the sale of second-hand religious items and explore the role of municipal sharing websites in diverting things from incinerators in Germany. They discuss why mothers in Denmark prefer the exchange and circulation of used children's clothes to buying new items and analyze activities undertaken to expand the lifespan of car tires in Lagos. They explore how American university campuses that are overflowing with used microwaves, fridges, or desk lamps tackle issues of waste.

When authors examine practices of reuse, repair, sharing, and care, they do not lose sight of larger social, political, cultural, and economic contexts where these activities unfold. They analyze global economic and power dynamics that structure the trade in used appliances, like TV sets, between the UK and Tanzania. They show the environmental hazards of working with used tires in urban Nigeria. They challenge the fashionable use of products made of "reclaimed wood" sourced from deserted homes in Detroit. Is this indeed an "admirable" sustainable practice or an act of greenwashing that unfolds on the backs of those evicted from the homes from which the wood was taken? Contributors question the ethics of donated clothes in the context of religious volunteer-tourism in the Dominican Republic. What is the point of forcing second-hand clothes on those who do not want them? Is reuse socially destructive in this context?

Theoretically, this volume contributes to debates about second-hand cultures and economies of reuse, repair, sharing, and care that engage themes of environmental sustainability, inequality, poverty, degrowth, just transition, and social justice across the globe. Second-hand cultures and economies of reuse, repair, sharing, and care are situated at complex intersections of necessary social and environmental transformations, applause, and recognition for worthwhile sustainable projects and activities, job opportunities, fields of dangerous labor, toxic exposure, and exploitative work relations. Few would debate the necessity of comprehensive economies of reuse, repair, sharing, and care, but how can this be achieved without harm to and exploitation of those who do the work? Can individuals in Germany or Denmark celebrate their "sustainable" use of second-hand goods when these goods are the products of exploitative capitalist production in Bangladesh or Vietnam? Can sustainable activities be applauded when reused or recycled goods, even after their prolonged lifespan, still end up in landfills or workshops in the Global South to be dismantled under harmful conditions?

Analyzing "positive" examples of second-hand cultures and economies of reuse, repair, sharing, and care, it is crucial to recognize how they are situated in global economic and political regimes. Is a "sustainable" product the result of exploitative processes somewhere across the globe that are invisible to consumers? What is the end-of-life reality of a thrifted shirt or a thrice-repaired toaster? How are we to understand the export and repair of TVs from the UK to Tanzania from a material and an ethical perspective? Without doubt, repairing goods and using second-hand items are worthwhile activities; however, it is paramount to ask questions about their sourcing, production, circulation, materials, toxicity, and end-of-life scenarios to understand their position in global networks and impact on people, animals, and the environment.

In addition to questions about sourcing, manufacturing, and end-of-life, it is vital to analyze how effective individualized practices can be in the face of a planetary crisis. Can individual actions and small-scale projects contribute to robust economic, social, and ecological change, and challenge capitalist hyper-consumption and ecological destruction? Can reusable coffee cups, repaired TVs, and thrifted jeans really make a difference?

Historical Notes on Reuse, Repair, Sharing, and Care

For most of human history, second-hand cultures and economies of reuse, repair, sharing, and care were integral to everyday life. People owned little and used what they had as long as possible. Frugality was a universal value. People reused, repurposed, and recycled things without using these terms. It was ordinary and shared practice (Alexander and Reno 2012, 5). Starting in the nineteenth century, modern industrial capitalism unsettled time-honored values and practices and offered ever more and "better" products to buy and subsequently discard the obsolete, outdated, or simply unwanted, now defined as garbage (Rogers 2005; Pellow 2004). From the beginning, the capitalist extraction of raw materials, production of goods, and disposal of waste were based on the exploitation of workers and degradation of the planet (Royte 2006; Miller 2000). In the nineteenth century, a powerful consumer culture emerged that marginalized earlier practices and economic circuits. Repair, once a core element of good stewardship, was slowly relegated to history or became a mark of poverty (Strasser 2000, 25). However, repair, reuse, repurposing, and care practices never disappeared (Alexander and Reno 2012). In the early twentieth century modern consumerism exemplified by fashion and new household appliances, bought in large and lavishly decorated department stores, became a core feature of distinction among urban elite and increasingly middle-class households (Zola 2012). At the same time, urban second-hand economies consolidated and catered to those excluded from the new consumerism.

At the turn of the twentieth century, thrift stores, flea markets, and other second-hand venues dotted cityscapes. They were rooted in traditions of rag pickers and bone or metal collectors who had long travelled towns and cities salvaging materials (Miller 2000). Old-style garbage collectors and second-hand dealers made cities cleaner, diverted materials from informal dumps, and created jobs for those who could not find other work (Lapolla 2009 [1935]). Their work was dangerous and precarious

(Pellow 2004). Emerging large charitable thrift stores consolidated used goods markets. In the United States, Salvation Army opened its first thrift store in 1897 and Goodwill in 1902 (Le Zotte 2017, 19). They sold used goods and aimed at "reforming" workers and customers (ibid., 17). They employed those deemed "unemployable," used volunteer labor and professionalized the used goods trade and marginalized small rag pickers and traders (ibid.).

In the face of mass production, rapidly changing patterns of consumption, and growing social inequality, many individuals and households continued to be producers, makers, and menders. By force or choice, they maintained reuse and repair practices and patronized second-hand markets (Strasser 2000; Alexander and Reno 2012). Frugality remained a value. Clothes continued to be downcycled from Sunday to work wear or reworked into children's clothing, and finally turned into cleaning rags or doormats.

After WWII, global capitalism solidified into a system of over-production and hyper-consumerism. Those with the means consumed ever more at ever-faster rates and produced unprecedented quantities of trash (Rathje and Murphy 2001). By the 1960s, consumerism and excess possessions filled middle class American homes to the brim and many needed to shed stuff to make room for more stuff. Until then, extra possessions had been "passed on through kinship and friendship networks" as "hand-me-downs" or given to charitable organizations for rummage sales (Crawford 2014, 25–26). As quantities grew, households experimented with new ways of ridding themselves of stuff. Hoping to make some profit from their excess belongings, people turned to sell things themselves, while having fun with their neighbors. Gretchen Herrmann (2019) dates the emergence of garage sales in the United States to the 1960s because of explosive consumerism. Herrmann and Stephen Soiffer explain that by the 1980s, garage sales represented a "cultural style, a quasi-ideological rejection of consumerism" (1984, 411) for consumers who started to value the "recycling of goods" and its environmental impact (ibid., 413). They had become an integral part of "mainstream American life" (Crawford 2014, 26).

In the late twentieth century, yard sales, or car-boot sales in the UK made the landscape of second-hand exchanges and other aspects of growing economies of reuse, repair, sharing, and care more visible in the Global North. They connected with older circuits and networks across the globe (Minter 2019). In the twenty-first century, despite the pressures of the global capitalist regime or because of them, people worldwide continue to make, repurpose, reuse, recirculate, share, and repair goods (Corwin 2018; Schmid 2020; Isenhour and Berry 2020). Driven by poverty, the necessity of austerity, wishes for frugality, and a heightened awareness of environmental crisis, people adopt practices and design creative projects to make ends meet, take care of each other, create jobs, and battle environmental degradation (Murphy 2017; Medina 2007; Carr 2023). Often it falls to the poor or immigrants to do dirty and dangerous resource recovery work (Gregson et al. 2016; Kuppinger 2023). Many projects and efforts remain local and invisible, but the results of their efforts are considerable.

Recent Economies of Reuse, Repair, Sharing, and Care in the Global North

Starting in the 1990s Nicky Gregson and Louise Crewe conducted groundbreaking research on second-hand economies in the UK. When most analysts focused on work and value in the formal capitalist economy, Gregson and Crewe studied economic exchanges and social interactions at car-boot sales and other venues and highlighted the economic, social, and cultural contributions of marginal second-hand economies (1998). They insisted, "that second-hand worlds matter" (2003, 1) and identified second-hand markets as sites of "resistant consumer practices" (1998, 39) that employ different rules (haggling for prices), pricing mechanisms, and value systems. These markets are embedded in complex fields of "finding, fixing, bartering, trading, and gifting and selling" and create value in unexpected spaces (Isenhour and Berry 2020, 298) that offer mundane economic transactions and "fun and sociality" (Gregson and Crewe 2003, 41) for sellers, buyers, and visitors (Herrmann 2006; Murphy 2017). "People go to thrift stores for pleasure" (Machado et al. 2019, 391) and many second-hand venues are meeting places where "peer-to-peer collaborative consumption fosters community" (Albinsson and Yasanthi Perera 2012, 305). Machado et al. argue that the more people go to streets fairs and similar events, the stronger their relationships become with people "in these circuits and the stronger their feelings of rejection for the traditional production chain" (2019, 303). This results in "financial benefits and critical behavior" (ibid.). Benedikt Schmid (2019) speaks of the "communality" of shared repairs spaces that offer cooperation and inclusion.

He insists that spaces like repair cafés or open repair workshops facilitate the "creation of (public) spaces for politics and debate" (ibid., 244). Chantel Carr links repair work to care for the world and people which are vital for "adapting to, and coping with the conditions of planetary breakdown" (2023, 223).

Second-hand markets offer goods at cheaper rates but also reflect, "the very same motivations that shape consumer culture more generally" (Gregson and Crewe 2003, 11). They also harbor "potential for consumer activism and empowerment" (ibid.) as participants play an "active role" (Machado et al. 2019, 319) and subvert dominant consumption patterns, when they prolong the lifespan of goods and avoid buying new things. Second-hand markets and related projects further "offer the potential to better educate individuals in the value of sustainability" and trigger "deeper engagement" with such issues (Murphy 2017, 170; Schmid 2019, 238). Gregson and Crewe point to the "birth of value" (2003, 2) in second-hand exchanges that departs from dominant economic practices, when "people create the conditions for value to emerge in the context of transactions" (ibid.). Sellers and buyers are agents and producers of value rather than passive consumers. They produce "commodity biographies" (ibid., 6; Appadurai 1986) and infuse meaning into moments and spaces of exchange. Selling and buying second-hand goods involves more time and labor than mainstream transactions, as sellers might repair items and customers search markets for desired goods.

Gregson and Crewe (2003) analyze the rapid proliferation of charity shops (charitable thrift stores) and other sites of second-hand exchanges (car-boot sales; auctions, retro/vintage shops, antique fairs, classifieds) in the UK since the 1970s. Some car-boot sales developed into fun fairs with food and bouncy houses as affordable family entertainment (ibid., 32). Urban charity shops professionalized and now line shopping miles, marking their arrival in the middle of society. Retro/vintage shops dot gentrifying and cool quarters (ibid., 39), indicating complex links of first and second-hand circuits and the commodification of useful/fashionable discards (ibid., 62; Isenhour and Berry 2020, 305). Here second-hand shopping becomes a sign of distinction and confers cultural capital/recognition to shoppers (Steward 2020). Gregson and Crewe understand much of the second-hand economy as sites of resistance and mindful and affordable consumption (1998, 40), but in the context of some retro/vintage shops, they point to limits. Such stores "produce second-hand worlds that are continually articulated with and understood in relation to spaces associated with and identified with 'the new'" (2003, 88). Curated second-hand boutiques resemble those that sell new products and function under the similar market rules, catering to particular social, cultural, and "taste communities" (ibid., 108).

Regardless of the organization of second-hand venues, Gregson and Crewe (2003) maintain that consumers carefully design their practices and understand them as politically, socially, and ecologically relevant. Interlocutors tell Gregson and Crewe that they practiced "green consumerism" (Gregson and Crewe 2003, 103). One person emphasized they had not bought new things in 20 years, another added that they only buy new underwear (ibid., 104), underlining beliefs that individual action matters. Another interlocutor remarked that buying used is "about fighting against the capitalist system. It's about not being drawn into consumerism" (ibid., 103). They insist on their agency and political participation via consumer choices, remake linear consumer models of production–consumption–discarding, and propose different "temporalities of possession and disposal" (ibid., 117), when they keep goods moving and reinsert them into circulation when they no longer need them (Bohlin 2018). Second-hand spaces offer transactions that "subvert and transgress the experience of the mall and high street" (Crewe and Gregson 1998, 40). Of course, there is always the end-of-life moment for goods, but it can be postponed with good stewardship, reuse, and mending, and repurposing.

The sourcing of used goods is an increasingly competitive process where savvy traders comb markets for hidden gems and make sizeable profits which creates "complex, but highly localized and seemingly infinite chains of second-hand commodity circulation and re-use" (Gregson and Crewe 2003, 131). Traders seek "to buy (from the 'unknowing') to sell (to the 'knowing') in ways that are clearly about capturing competitive advantage and profit" (ibid.). In the process, valuable objects are removed from disadvantaged, often rural, communities to profitable urban markets (Isenhour and Berry, 2020). Second-hand markets become complex circuits that require considerable "commodity and geographical knowledge about disposal and sourcing" (Gregson and Crewe 2003, 141) and reflect dominant capitalist market dynamics and patterns of exploitation and exclusion.

Regardless of problems and limitations, many second-hand shoppers, makers, tinkers, and menders transfer considerable value from the capitalist economy to more socially and ecologically minded and historically grounded resilient economies of reuse and care (Graham-Gibson 2008).

They create "spaces of alternative economizing that are partially removed from capitalist valorization" and are "genuinely subversive" as they challenge the "apparent verities and certain singularities of capitalism" (Schmid 2019, 231, 247). Theorizing the proliferation of experimental social and economic projects, Graham-Gibson point to the global infrastructure of meetings, websites, and other media that describe, support, and solidify their impact (ibid., 614). They speak of "world-makers," who in small groups tackle humanity's challenges (ibid.). Graham-Gibson insist that invisible activities and projects are connected in their struggle for alternative futures, just economies, and planetary health. They state that non-market transactions and unpaid household and care work account for 30–50 percent of economic activities in countries across the world (ibid., 615). Current economies of reuse, repair, sharing, and care unfold in such contexts. Garage sales, repair shops, swapping events, and repurposing activities are "spaces of difference" (ibid.) where patterns of labor, valuation, exchange, and sociality differ from capitalist markets and seek to "contribute to social well-being worldwide" (Graham-Gibson 2008, 615). This global care/alternative economy illustrates "the possibility of new economic becomings, rather than a condition of their impossibility" (ibid., 619).

Benedikt Schmid (2019) insists on the importance of repair and maintenance in more circular economies. He situates "repair within the context of a broader movement that work [sic] towards 'repairing the world'" (ibid., 234). Repair responds to visible breakdown that disrupts the flow of life and smooth functioning of consumer society (Tironi 2019). Discarding a broken object and buying a new one is the "best" solution in a capitalist society. Repair challenges this desired trajectory. Schmid argues that repair practices are "counter-hegemonic movements with the potential to disrupt the stories and trajectories of waste, consumption and growth" (2019, 234). Whether by financial necessity or choice, repair and maintenance are crucial in the making of futures marked by care and sustainability (Corwin and Gidwani 2021). Repairing or repurposing are robust elements of resilient parallel economic networks. Schmid concludes that repair is "rebellious" (Schmid 2019, 239) and deeply interwoven with activities that similarly aim to "reconfigure commodity chains" and challenge linear economies (ibid., 242).

Analyzing the impact of a pair of thrifted jeans, a repaired vacuum cleaner, or an hour of work in a garden, one should not measure the dent these activities make in the capitalist economy. Instead, these are worthy additions to time-honored global economies of reuse, repair, sharing, and care that represent conscious stewardship and creative production (Kuppinger 2021, 2023). Focusing on small acts frees us from the grasp of dominant discourses and theorizing that posits individuals as powerless (Graham-Gibson 2008). Small projects and spaces seed more robust and long-term transformations (Franck and Stevens 2007; Hou 2010) and represent "mundane forms of power" (Graham-Gibson 2008, 619) that reside in ordinary practices. Shaping messy lifeworlds and practicing care and stewardship foster "a love of the world" (Hannah Arendt, quoted in ibid.) and "draw on the pleasures of friendliness, trust, and companionable connection" (ibid.). Graham-Gibson challenge the "erasure" of small efforts by neoliberal discourses and highlight their impact and connectivity. It is paramount to map geographies of social and ecological practices and noncapitalist transactions, demonstrate their global connectivity and impact, and highlight integral elements of love, care, cooperation, and trust. Small efforts "create geographies of collective responsibility" (ibid., 622), maintain local knowledge, and engender moments and spaces of shared humanity. They foster the unexpected, individual agency, creativity, community, and challenge capitalism (Crewe and Gregson 1998, 50).

Joe Smith and Petr Jehlička (2013) point to the "quiet sustainability" of people in Poland and Czechia who tend to their gardens, grow healthy foods, take care of the soil, gift their produce to others, and share their "joy, exuberance, generosity, care, and skills" (ibid., 156). Smith and Jehlička highlight the "daily practice of a satisfying life" that does not loudly announce its sustainable achievements. Urban farms and community gardens (Ableman 2016; Hou 2014; Milbourne 2010) often include marginalized constituencies (Mares and Peña 2010; Truitt 2012). They provide nutritious produce for many, create urban oases in dense neighborhoods, and produce communal cohesion and networks, often across generational lines (Sokolovsky 2018; Allen & Charles 2013). For participants doing is enough, much like the thousands who have quietly but deliberately inserted "upcycling, reducing, reusing, recycling, repairing, and redistributing possession through sharing, donating, and ridding" into their lives (Albinsson and Yasanthi Perera 2012, 308; Milbourne 2010). Invisible acts like "[C]rafts that take place in the shed, at the kitchen table or the village hall" outline "quietly revolutionary and ethically sustainable versions of how we might live and work: our values, relationships with others and the environment" (Hackney et al. 2020, 39; Corbett 2024). Small spaces and "households are key sites within which the challenges of environmental sustainability and climate are negotiated and material made sense of" and deserve more attention (Carr 2023, 233). Their transformative potential is considerable. A look at popular texts about individual projects like zero-waste lifestyles, minimalist living, or plastic-free living further illustrates this point.

In 2006, Bea Johnson and her family of four embarked on a low waste, and eventually zero waste journey in California. Johnson (2013) chronicles her zero-waste experience and advises interested readers offering recipes for zero waste cosmetics, cleaners, and food. She was lucky to have a store that sold bulk and unpackaged goods and accommodated her quest, and to be financially able to shop at this store. Whether or not zero-waste is doable, Johnson is a veteran of a movement that inspired thousands and helped them live low-impact lives. Beth Terry (2012) had a plastic "epiphany" after she saw images of a bird that died with a stomach full of plastic. She went cold turkey with plastic and tried to shed all plastic in her daily life. She introduces individuals who pursued similar practices and goals, like Jean Hill who at the age of 85 fought for a plastic bottle ban in her Massachusetts town (ibid., 102–103). Terry insists on the relevance of networks, meeting similarly minded people, and joining political activism to increase the impact of emerging care practices. On a random Saturday, Joshua Becker (2016) had a stuff "epiphany," when he tried to clean his overstuffed garage and never found time to play with his son because there was too much stuff to deal with. Becker and his family decided to rid themselves of many/most burdensome possessions and embarked on a journey of minimalist living. They became forerunners of the minimalist movement (D'Avella 2015). Johnson, Terry, and Becker describe individual radical lifestyle changes that resulted from their experiences or reflection. They were horrified by their own hyper-consumption as privileged individuals in the Global North and the consequences for people, animals, and the planet. They chose individual practices to address global problems and inspired others to adopt similar practices.

Global/ized Economies of Reuse, Repair, Sharing, and Care

Across the globe, millions perform remarkable social and ecological care work. Their efforts contribute to communal well-being and planetary health, but are rarely recognized, or are even shunned by those around them (Fredericks 2018; Millar 2018). Chantel Carr points to the "invisible bodies that will inevitably hold our world together" with their "work, skill and value across all sectors" (2023, 224). The waste collectors in Cairo, Egypt, who are pioneers of reuse, recycling, and food salvaging, are one such example.

In the nineteenth century, what little waste there was in Cairo was sold to the makers of *fuul* (bean dish cooked overnight) or public bathhouses (Assad and Garas 1993 [1994], 2). The advent of modern consumer culture in the early twentieth century altered types and quantities of waste, and rural migrants started to collect refuse from wealthier Cairo households and continued to sell it to those who needed fuel (Volpi 1996, 14). They made a living, kept the city relatively clean, and established networks of reuse, repurposing, and recycling. In the mid-twentieth century, more rural migrants (mostly Christians from Upper Egypt), joined these efforts and subcontracted with the earlier collectors. The newcomers handled the street collection of trash; they became known as *Zabbaleen* (garbage workers). They used organic waste (most of the refuse) for their pigs (earlier Muslim collectors were prohibited by their religion from keeping pigs; Fahmi 2005, 156).

In the 1950s, the two groups set up settlements on the urban fringes (Kamel 1994) and extended their services in the growing city that did not have a municipal waste management system (Haynes and El-Hakim 1979, 104). Garbage proliferated and the Zabbaleen improved their sorting and recycling system. Materials were sold to dealers who resold them to workshops for reuse or recycling (ibid., 103). The Zabbaleen devised organic solutions and tailored work regimes and techniques to consumer patterns and resource markets. Waste provided work for thousands and recycling workshops emerged in Zabbaleen communities, where, for example, plastic was recycled into dishes and aluminum into cutlery. The Zabbaleen provided a tremendous service to the city and environment but remained poor and struggled with alarming health problems (Haynes and El-Hakim 1979, 105; Fahmi 2005, 157). In 1979, one Zabbaleen communities had an infant mortality rate of 40 percent (Haynes and El-Hakim 1979, 105).

In the 1970s, the Zabbaleen organized to create a cleaner work environment and on-site recycling that offered more profitable jobs. Their new association (in the Muqattam Mountain community) set up a composting plant for pig manure and organic waste (Kamel 1994) and a rag-recycling unit, that included social and health services (Kamel 1994, 118). A recycling school for children and youth followed (Iskander 2009). By the 1990s, the Zabbaleen operated facilities for the recycling or reuse of aluminum, plastic, cloth, paper, cardboard, tin, bones, glass, and organic waste (Kamel 2009) but continued to live on the margins of urban society. At the same time, the municipality sought to replace them with international waste management companies. Regardless, in the late 1990s the Zabbaleen handled about 30–40 percent of Cairo's garbage, approximately 3000 tons/day (of a total of about 9000–10,000 tons, produced by 14 million residents; Aziz 2004, 9; Kamel 2009). They had developed a unique waste management system with an exemplary recycling rate of 80–85 percent (Boyd 2008, 47; El-Dorghamy 2009). The city tried to contract waste management firms that were asked to recycle a mere 20 percent of the collected waste (Fahmi and Sutton 2010; Rashed 2002; Kamel 2009). The Zabbaleen weathered numerous setbacks (the culling of their pigs in 2009 in the politicized context of the global swine flu) but kept bouncing back. In 2009, they processed 6000–7000 tons of waste daily Slackman 2009; Williams 2009).

Cairo's Zabbaleen represent an incredibly efficient waste management, reuse, and recycling system that organically grew out of pre-industrial practices. For decades, workers improved workflows and techniques as consumption and refuse patterns changed. The example of the Zabbaleen also illustrates the high human costs of such work, especially in the years of rapid growth of garbage in the mid to late twentieth century. Families joined this work for lack of other options. As the community, housing, and work conditions improved, many stayed for the independence this work offered and the cohesion of the community (Iskander 2009, for similar phenomena in Brazil and India, see Millar 2018; Saraf 2020). Risks and hazards, albeit in a reduced manner, remain perilous features of the Zabbaleen's lives.

The Zabbaleen's work demonstrates the complexity of recycling economies. It is locally developed, but global in scope. Products that end in the Zabbaleen's sorting grounds are globally produced. Global corporations make profits from the soda or shampoo bottles the Zabbaleen sort. Even if locally manufactured, profits return to global capitalist centers. The Zabbaleen channel tons of materials into recycling streams, save vast quantities of CO_2 emissions, and eliminate toxic leachates from landfills. They are part of global recycling networks, and some retrieved materials like PET bottles are traded in global markets (Furniss 2014). The Zabbaleen save millions of tons of waste from landfills, contribute to a cleaner city, and recirculate valuable resources. The case of the Zabbaleen illustrates historical, social, cultural, religious, and political elements that are rooted in preindustrial practices and, over more than a century, interconnected with global capitalism to create a finely tuned reuse and recycling economy.

Adam Minter (2019) travels around the world and reports about second-hand activities and their global connections. He visited professional declutterers in Japan who empty apartments of the deceased when families are not able to do this demanding labor. While much of their possessions end up in the garbage, some feed into Japan's vast second-hand industry including auctions, brick and mortar stores, or online shops. With the abundance of used goods, companies specialize in the export of Japanese second-hand items overseas turning domestic circuits into global ones. Minter notes that the "high price of purging personal belongings is one factor driving reuse in Japan. The other is globalized secondhand" (2019, 87). Minter's interlocutors explain that people in the Philippines "have a deep love for Japanese products" (ibid.). One company exports Japanese second-hand items to 40 countries (2019, 91) and made about $ 1 billion in 2010 (2019, 93). In their warehouses, washing machines stand ready to go to Vietnam and knitting machine are destined for Nigeria and Pakistan (ibid.). Company representatives note that affluence, not poverty, drives this economy as ever larger groups in emerging economies like the Philippines, Malaysia, or Vietnam want appliances and other consumer goods (2019, 96). Second-hand goods are their entry to the universe of globalized consumption.

Minter chronicles the vast small-scale second-hand trade between the United States and Mexico. Unlike the large containers shipped overseas from Japan, this trade is done by single pickup trucks that cross the border packed with goods. One trader notes that mattresses, appliances, and clothes are the best-selling items (2019, 110). Thousands of small traders "evacuate the unwanted stuff generated across the Southwestern United States to Mexico's up-and-coming consumers" (2019, 108). Minter describes the enormous volume of stuff that daily crosses the border to Mexico and concludes that this is "one of North America's most environmentally sustainable businesses, but nobody notices unless they're in it" (ibid.). Minter explains that two-thirds of the merchandise at Goodwill stores in Arizona never sell and is moved to the Goodwill Outlet Center where clothing, hard goods, and glass items are sold by the pound (e.g., 29 cents for a pound of glassware; 2019, 118). The Outlet manager estimates that 80–90 percent of their customers are Mexicans who resell things on the other side of the border (ibid., 119). Clothing that does not sell in the Outlet will be exported to countries across the world. Most hard goods end up in landfills (ibid., 123).

The devastating effects of fast fashion and resulting global second-hand circuits are well documented (Cline 2013; Thomas 2019; Hansen 2000). The United States, the UK, Germany, South Korea, and China are the world's largest exporters of second-hand clothing (Brooks 2019, 141). Circuits of second-hand clothing are complex and ever changing as countries ban them or demand high import duties. Nigeria placed strict regulations on used cars, computers, TVs, and clothes (ibid.). Minter describes a multi-billion dollar business in used items between Benin and Nigeria where savvy Nigerians buy used goods and manage to get them into Nigeria at cheaper rates (ibid., 143). Many Nigerians prefer used quality goods to cheap Chinese imports (2019, 144; see Ntapanta this volume). Consumers in Africa demand quality, as one trader stated: "We don't want garbage. We want fashion. We want quality. Not your garbage" (2019, 151). The final destination for clothes that nobody wants are landfills that burden local people with the excesses of global overconsumption producing, among other issues, water pollution when dyes and chemicals that leak into groundwater or rivers (Nkatha 2023). This problem is not restricted to Africa. The infamous illegal clothing dump in the Atacama Desert in Chile (the world's third largest importer of used textiles) is so large that is it visible from space (Johnson 2024).

Minter visits a rag factory in Ohio that converts unsold low-quality clothing into billions of cleaning rags, mostly for gas and oil industries where endless pipes and valves need to be wiped. The low quality and mixed fibers of fast fashion even affects the rag industries. One company, for lack of appropriate used materials, started to produce 100 percent cotton rag from new cotton which have a larger environmental impact than recycled ones (2019, 163).

In Ghana, Minter explores the second-hand trade with clothes, appliances, electronics, furniture, and bicycles. He notes that second-hand dealers outnumber those selling new items. In regional market cities like Tamale, Minter estimates a ratio of 100 to 1 of used versus new good sellers (2019, 243). Quality, long-lasting, and repairable products are in demand in Tamale and beyond. Young people in search of laptops prefer quality second-hand items to new Chinese products (ibid.). Minter examines the vast urban scenery of used electronics and repair shops, especially for TVs. In Ghanaian and other West African cities, TV-repair shops "are more common than Starbucks in Manhattan" (2019, 218). They repair tens of thousands of TVs (mostly from Europe, the United States, and East Asia) every week (ibid.). Their rate of success, reuse and resource conservation is remarkable and beats highflying reuse goals of recycling projects in wealthy countries. Local experts insist they can repair almost anything except when they lack parts. Similar things happen with imported used cars in the region (Minter 2019, 224). The second-hand trade and repair business create thousands of livelihoods and are dynamic, creative, and productive circuits where almost anything has value. Julia Corwin similarly notes that the labor of e-waste repair and processing workers in Delhi constitutes a "powerful sources of value" (2018, 14). Sellers, waste workers, and repair experts across the Global South provide impressive services to consumers in wealthy countries, local customers, and the planet. Vast reuse industries employ extraordinary skills to repair, but often also to "reinvent electronic commodities" (ibid., 19; see Harper 1987) as they convert e-waste from "rubbish to a valuable product" (Corwin 2018). Their work represents a "disruptive interaction with capitalist regimes of value" (Thoburn, quoted ibid., 22). The question of hazardous working conditions urgently remains when workers do good for people and the planet at the expense of their health.

Minter's final stop in Ghana is Agbogbloshie, often named "the world's largest digital dump" (2019, 249) where tons of e-waste and cars are dumped every year. Concrete estimates of the dump's volume vary and much sensational reporting has been done about it, but the fact remains that it endangers workers, local communities, and the environment. Selling mostly car parts, Agbogbloshie, is not a safe workspace, because "safety equipment is nonexistent, and the air is riven with the smell of burning plastic" (2019, 251). Agbogbloshie is the final resting place for used imports to Ghana, that are "used, repaired, and reused, often for decades" (2019, 253). Ghanaian consumers and repair experts perform stunning services to the planet extracting maximum use from products that previous owners in wealthy countries defined as waste. They make things more sustainable but pay the price for having the (toxic) remnants dumped in their backyard with "devastating consequences on the environment and human health" (2019, 254). Forty-thousand people live in Agbogbloshie, work with, and suffer from global consumerism. There are endless small businesses making and selling retrieved, repurposed, and recycled items, like stoves made from scrap aluminum or refurbished auto parts. Businesses make "new" computers from old parts. Minter calls these "thirdhand markets" (2019, 254). Wealthy countries and their consumers have much to learn from Ghanaian traders, repair experts, and consumers about productive reuse, repair, and sustainable practices. Their practices are remarkable, but workers face extraordinary health and environmental risks that reflect a horrendous injustice.

Experts in Ghana and elsewhere struggle with restricted access to repair details for cars, appliances, and electronics like cell phones (Houston 2019, 131). For example, "Apple refuses to make service manuals or parts available to independent repair shops or consumers" (Minter 2019, 227). "Anti-repair practices" greatly hinder the work of global repair specialists and increase the volume of electronic and other waste (ibid.), but also produced projects like iFixit, a company in the United States that creates repair manuals for products and posts them on the Internet. Minter reports that iFixit published a stunning 38,000 repair manuals. Kyle Wiens, the co-founder and CEO of iFixit noted that their website has over 100 million visitors a year (2019, 231). The company's goal is to change consumer behavior and to show that "repair is a viable option for people" (2019, 232). Like his colleagues in Ghana, Wiens emphasizes that getting good parts is fundamental to successful work. Wiens and other activists demand a right-to-repair and that companies must make repair information and parts available to all. In the absence of such support, informal cyber networks have emerged where repair experts exchange knowledge across the globe. Lara Houston (2019) writes about cell phone repair technicians in Uganda who participate in global repair infrastructures where repairers share information and software (ibid.). Locally such knowledge is disseminated via "apprentice relations" and constantly extended with "trans-local knowledge ... accessed via the Internet" (ibid., 135). She reports about networks where repair workers in Kampala exchange information with colleagues in "Denmark, India, Pakistan, Kashmir, Dubai, Poland, and Turkey" (ibid., 155).

Global Reuse and Care

Contemporary cultures and economies of reuse, repair, sharing, and care are intricately interwoven across the globe. Like never before, product biographies connect people and places across the globe via supply chains, use, reuse, and waste management networks. Egyptian waste workers sort French or American brand name shampoo bottles. International Danish mothers buy used clothing for their infants, which might have been produced in Vietnam or Turkey. Rag producers in the United States complain about the low quality fast fashion fabrics imported from China or Bangladesh that are not even good enough to clean an oily valve. Decluttered washing and knitting machines from the estates of elderly Japanese end up in upwardly mobile households in the Philippines, Nigeria, or Pakistan. Discarded mattresses and appliances are transported in pick-up trucks from the United States to consumers in Mexico. Cell phone repair specialist in Uganda routinely communicate with colleagues in Poland, Dubai, and Pakistan to solve repair challenges. Women in India turn worn out saris into global designer fashion.

Connections are limitless and crisscross the globe in intricate patterns. For the analysis of current economies of reuse, repair, sharing, and care, it is paramount to recognize these connections as well as their productive and destructive, exploitative, and dangerous features. Reuse and care are parts of a global material landscape that is characterized by unspeakable privilege and horrendous marginalization and exploitation. One can see the aesthetics of an upcycled wooden table sourced from "deserted" homes in Detroit *and* ask critical questions about this sourcing and who was forced out of homes in the first place. One can acknowledge the achievements of local reuse economies *and* ask about why some individuals are not deemed "worthy" of access to their goods. One can recognize the efficiency of workers in Lagos who maintain and reuse tires *and* ask questions about their health and the overall injustice of being recipients of used products from wealthy individuals and countries.

Ordinary people's work and everyday practices unfold on the background of global capitalist production but also reflect local wisdom, needs, and positions versus the global. Saving a tire in Nigeria, a sari in India, reusing a jam jar in the United States, and buying second-hand baby clothes in Denmark are connected activities that seek to create a better world. While motivations and concrete circumstances differ markedly, they are all acts of care for people and the planet. These activities improve the world and need to be recognized. They challenge dominant linear patterns of hyper-consumption, and emphasize values of frugality and respect for things. This does not mean that activities are perfect or beyond criticism and improvement. They are a start because they recognize problems and attempt to solve them. The challenge is to allow all "world-makers" to do so under the best possible and just circumstances instead of under pressure of poverty and marginalization. The chapters in this book introduce some such world-makers.

Chapter Overview

Chapters in this volume present ethnographic studies of how ordinary people across the globe live, revive, create, and refine practices of reuse, repair, sharing, and care as they try to make a living, seek to prolong the lifespan of things, create communities of care, seek to contribute to planetary health, generate awareness, and try to teach others. Authors ask critical questions about how some "sustainable" practices can harm individuals and groups.

The book includes four topical sections: Part I addresses spaces and practices of second-hand cultures and economies or reuse and care. Authors introduce contexts and spaces like yard sales, thrifts stores, a farmers market, a university composting site, and an urban sharing website and examine practices like passing used items along with their stories or teaching reuse practices at a farmers market or on a university campus. Part II explores repair practices, concretely the landscape of second-hand TVs and their repair in Tanzania. Chapters in Part III focus on the reuse of materials like clothes, tires, and wood. Part IV discusses religiously marked or inspired second-hand circuits like the resale of "sacred" objects, and the donation of clothing and other materials in the context of missionary trips.

Gretchen Herrmann examines resale and reuse circuits in upstate New York. She introduces the ReUse Center, a nonprofit thrift store, its daily work, and contributions to the community. She discusses private garage sales and juxtaposes the working of these two venues in the larger reuse and developing circular economy. She explores their social functions and socialities. Garage sales allow sellers to tell stories about their possessions and create ongoing narratives about things. Garage sales are pleasant ephemeral social spaces where items are passed on among neighbors and friends. Brick and mortar thrift stores sell and recirculate items but rarely tell their stories. A new owner can only imagine a story for the acquired second-hand goods. In contrast to temporary garage sales, thrift stores, because of their permanence and size, provide volume and predictability for customers. Stores are always open to their diverse clientele and offer a constant and sizable selection of items. They contribute to the community via social services and programming.

In Chap. 3, Brieanne Berry, Cindy Isenhour, Andrew Crawley, and Sujan Chakraborty examine the role of social capital in second-hand markets in Maine. They point to the ambivalent use of the concept, that on the one hand has been hailed as positive in the making of relationships and communities, but on the other hand is identified by some theorists as exclusionary and even elitist. Drawing on a statewide survey and ethnographic data, Berry et al. provide situated knowledge to analyze the role of social capital or more precisely social networks and relationships in moments of inclusion or denial of help in second-hand venues. Looking at "things in motion" and those who give, sell, buy, or are gifted used objects, they find relationships of trust among those involved (personal relationships, social network support, civic engagement, trust and cooperative norms). This resonates with more positive definitions of social capital, but the authors insist such data needs to be supplemented with ethnographic observation of concrete contexts. They note we need to analyze "the movement of goods through networks structured by relationships." Observing activities and interviewing participants at different second-hand venues, Berry et al. provide nuanced data especially about a sense of exclusion among some participants. Marginalized individuals are kept from the "relational world of reuse," being deemed undeserving and made to cross humiliating barriers of having to establish their trustworthiness. Some individuals' emergencies (a burned down home) are more easily recognized as "worthy" of free used items than others' (e.g. mental health, long-term housing status). The authors conclude for one shop "the strong bonds that link some to this shop keep others out" which rings true for larger reuse circuits. They emphasize the positive communal aspects of reuse economies but point to elements of exclusion that illustrate how well meant, often voluntary activities, are entangled in existing inequalities.

Susan Andreatta examines the role of farmers markets in the making of more sustainable communities in Chap. 4. Based on research in Greensboro, NC, Andreatta analyzes practices at the market that limit single-use items, showcase sustainable projects beyond the market, and encourage visitors to adopt sustainable practices. Analyzing the market through the theoretical lens of degrowth, Andreatta probes into market practices and how vendors and customers challenge hyper-consumerism. She examines efforts to promote the use of reusable plant and produce containers, jam and honey jars, and shopping bags. Market organizers are committed to sustainability and reuse beyond market days and highlight sustainable practices to visitors. Andreatta mentions the quarterly "Curiosities on the Curb" day, a second-hand and vintage event that highlights how to give used things a longer lifespan and provides environmental education by way of exhibits by organizations like a recycling center, cycling clubs, or solar panel companies. Andreatta interweaves market efforts with individual practices of her students and how they saved coffee cups or other plastic items with small changes propagated at the market. Andreatta concludes that mindfulness, as expressed in practices like reducing waste or reusing things is a crucial element in larger transformations that produce tangible greening effects.

In Chap. 5, Angela Storey and Henrietta Ransdell explore spaces of reuse on the campus of the University of Louisville, KY. They describe four projects: a free store, a food recovery network, a campus garden, and a composting program. Largely run by students, these projects create spaces of material reuse and care and are sites of cooperation and relationships. Storey and Ransdell argue that in our current "wastocene" era, it is crucial to provide spaces of environmental learning. Exchanging clothes and other items, recovering food, working in a campus garden, and helping in a composting project highlight reuse practices and create communities and meaning. Material care translates into care for each other and shows how participants can remake the world. The authors highlight the effectiveness of experiential learning, and how concrete practices are political, empower participants, and help students understand that they are part of something larger and can make a difference.

Petra Kuppinger (Chap. 6) examines urban landscapes of reuse, repair, sharing, and care in Stuttgart, Germany. Analyzing a repair café, thrift store, and municipal gifting website, she asks whether these projects are individualized efforts, or add up to a more effective field that produces tangible results and contributes to long-term transformations. Analyzing repair processes and social encounters at a repair café, the reception, handling and sale of used items in a thrift store, and free exchange of goods via a municipal website, Kuppinger introduces a robust urban geography of reuse and care. This landscape is supported by individuals who have things repaired even when this means spending a Saturday morning at a community center, spending time at thrift stores, or on free exchange websites to locate suitable second-hand items. These activities are not only about saving money. Why else would somebody travel across the city to pick up a small lamp or four cups that they found on a free exchange website? Kuppinger argues it is crucial to identify these individual practices, their social and ecological motivations, and recognize the robust networks of reuse and care that are solidified with these acts.

In Chap. 7, Samwel Ntapanta examines the role of second-hand TV-sets in Tanzania. For many people repair, reparability, and the availability of skilled repair people play crucial roles in the purchase of electronics. Ntapanta notes that many consumers prefer used popular brand, especially Japanese products, to cheaper new ones (mostly Chinese). Using the example of his family's purchase of a TV-set, he shows how his father went to the market with a specific brand in mind, because it had a great reputation, but crucially, because there were skilled repair workers available in his region. Ntapanta insists that dominant European and Northern American perceptions of African countries as destinations of electronic waste need to be revised, as they do not consider the vibrant trade in second-hand goods and remarkable repair skills available for such products. A TV-set that enters Tanzania might be a long way from a landfill and will be mined for parts before the remaining materials ever hit a landfill. Regardless of this prolonged lifespan of electronics, Ntapanta points to ongoing circuits of inequality and marginalization manifested in the export of electronic "waste" to Tanzania.

Erika Kuever (Chap. 8) introduces the circuits of second-hand children's items among (new) mothers in Denmark. Danish municipalities organize groups for new mothers (and fathers). Kuever introduces such a group that includes international mothers, which helps them maneuver new motherhood in general and initiate them into local parenting practices. Being confronted with a new climate and different expectations about childhood experiences (much outdoor life despite a cold climate), the new mothers explore material contexts of childhood. Using second-hand items (*genbrug*) is a crucial value in these groups and Danish society, tied to concerns about environmental degradation, waste reduction, and ideas of circularity. Group members embrace these values, based on different individual motivations, including high prices for new quality goods, moral values of reuse, sustainability concerns, and community-building elements of private networks or local thrift stores. Kuever highlights dynamics of "things in motion," what they do for momentary owners or stewards, the community, and the planet. Kuever's sample of international mothers is biased toward educated middle-class individuals who could afford new items but chose not to, which sets them apart from less financially privileged families. The international mothers, almost without exception, adopt *genbrug* practices, which they note provide joy, satisfaction, and a sense of community. Kuever explains that in follow-up interviews when the children were three years old, some families had abandoned second-hand practices when it became harder to locate good items for older children. Those most dedicated to environmental issues continued their reliance on second-hand circuits.

In Chap. 9, Simon Coleman, David Garbin and Andrew Esiebo analyze the landscape of imported second-hand and end-of life tires in Lagos, Nigeria. In their photo-essay, they show how second-hand tires are used and repaired as long as possible, recycled into fuel, furniture, shoes, automobile parts, pieces of art, or used to build asphalt roads. There are "vulcanizers" who fix tires for car owners to extract maximum use from old tires. These roadside experts battle assumptions about their work and seek to redefine it as skilled and essential roadside labor that keeps the city moving. Air becomes material here and needs to be kept sealed in to guarantee urban mobility. In Lagos, the lives of tires extend far beyond their original purpose and a fluid exchange system directs tires to different reuses. The authors introduce an artist who makes furniture from tires, appreciates their longevity, and teaches young people her techniques. Tires can be used during the raining season to create pathways through flooded streets or for children to play. Coleman, Garbin, and Esiebo situate tires as relevant objects in a cityscape were global economic processes intersect with local conditions, needs, and aesthetics.

Subhadra Channa examines the complex landscape of the reuse and recycling of saris in India. In Chap. 10, she explains that saris are not only six yards of cloth, but objects of social transactions, individual empowerment, and creativity, and a source of income and status. Creatively reused and recycled, saris can provide recognition, respect, and important social relations to their owners. Describing time-honored practices of frugality, Channa illustrates the many ways of reusing and recycling saris. Saris are downcycled in households and made into children's clothes, diapers, or cleaning rags. Channa describes the practices of *kantha* whereby worn-out saris are layered and stitched together into new materials and products. In recent years, this traditional recycling method has been discovered by global fashion circuits and some *kantha* products now fetch considerable prices. *Kantha* work that is produced for global markets provides some income and empowerment to marginalized women, but Channa cautions against blindly celebrating these recycling processes as they unfold on the back of economically marginalized women who see little of the profits made with their labor.

In Chap. 11, Simon Johansson asks uncomfortable questions about "sustainable" furniture produced from reclaimed wood in Detroit, MI. Encountering a beautifully upcycled dining table at friend's house while conducting fieldwork on urban regeneration and gentrification, Johansson explores the production of such upcycled furniture from wood sourced in abandoned homes in the city. Pieces like Johansson's friend's table decorate the homes of some young white upper-middle class returnees to the city. They might proudly display these expensive pieces as symbols for their claim to the city or links to their family's history in the city before they left for the suburbs decades ago. Johansson analyzes the uncritical celebration of circular economies and asks pressing questions about the context of materials' sourcing. He notes that economically marginalized households of color might have been the victim of eviction and their houses subsequently become the source of wood for upcycling.

Nicole Nathan (Chap. 12) explores role of second-hand clothing donations in American evangelical volunteer-tourism to the Dominican Republic. Missionary volunteer tourism is advertised as transformative experiences for the travelers. Less thought is given to the recipients and practices of "donation dumping." Missionary groups insist on giving second-hand goods over cash; a practice rooted in colonialist ideologies of donors knowing best what is good for recipients. These gifts are often unwanted, ill conceived, or even damaged and do little to improve recipients' lives. Some goods offend people's dignity, right to choose, and cause costs for import duties and disposal. On a national level, donations harm local economies and decrease local production and demand, and intensify global inequalities. Marginalized Dominicans receive discards of the wealthy, which were produced by similarly marginalized individuals elsewhere in the Global South. Second-hand gifts that show the traces of use, Nathan argues, underline the power differences between volunteer tourists and Dominicans. Givers feel good about their Christian acts and good "use" of things (and create room to buy more for themselves). Second-hand goods might be ecologically sustainable but are socially unsustainable when they deny agency to recipients and hurt local economies. Nathan points to the irony of CAFTA-DR free trade agreement with the U.S., which resulted in the decline of Dominican textile exports, and Dominicans becoming targets of donation dumping. Equally concerning is that American volunteer tourist spend most of their money (airfare, lodging, food) on themselves. Nathan concludes that donation dumping is an unsustainable second-hand circuit that fosters more consumption for the wealthy and burdens the poor with the discards of the rich.

In Chap. 13 James Bielo examines the second-hand market for Christian material culture items that unfolds in brick-and-mortar shops and on the

Internet and raises ethical questions about both sourcing and selling. For believers, a core question is whether it is ethical to sell what is coded as sacred, for profit on the capitalist market. Who provides guidance and possible limits for transactions in this market? Bielo introduces exemplary objects, resellers, buyers, and transactions and illustrates their positionalities and probes into motives of resellers and buyers. Some resellers make a living off the commodification of sacred objects. Others sell to ensure that these objects remain in the "right" hands after they were discarded in estate or church sales. Bielo notes that Christian material culture items circulate in a consumer society like other objects, yet many participants in this trade insist on its different ethical framing as they curate special collections or salvage sacred object from the capitalist market. Whether religious or not, resellers share an ambivalence about dealing with "sacred waste" that might merit different treatment from profane discards. Bielo argues that a certain enchantment pervades the merchandise that sets it apart from profane items regardless of the religiosity of those involved. He concludes that even though situated in the capitalist market, second-hand Christian material culture items tend to maintain their sacred qualities, and these might increase with exchanges.

References

Ableman, Michael. 2016. *Street Farm*. White River Junction: Chelsea Green Publishing.

Albinsson, Pia, and B. Yasanthi Perera. 2012. Alternative Marketplaces in the 21st Century: Building Community through Sharing Events. *Journal of Consumer Behaviour* 11:303–315.

Alexander, Catherine, and Joshua Reno. 2012. Introduction. In *Economies of Recycling*, ed. C. Alexander and J. Reno, 1–34. London: Zed Books.

Allen, Will, and Charles Wilson. 2013. *The Good Food Revolution*. New York: Avery.

Appadurai, Arjun. 1986. *The Social Life of Things*. Cambridge: Cambridge University Press.

Assad, Marie, and Nadra Garas. 1993 [1994]. *Experiments in Community Development in a Zabbaleen Settlement. Cairo Papers in Social Science* 16(4) Winter.

Aziz, H. 2004. *Improving the Livelihood of Child Waste Pickers: Experiences with the 'Zabbaleen' in Cairo, Egypt*. Gouda, Netherlands: WASTE.

Becker, Joshua. 2016. *The More of Less*. New York: Waterbrook Press.

Bohlin, Anna. 2018. "It Will Keep Circulating": Loving and Letting go of Things in Swedish Second-Hand Markets. *Worldwide Waste: Journal of Interdisciplinary Studies* 2 (1): 3.

Boyd, J. 2008. The Garbage People of Cairo. *Geographical* 80:44–48.

Brooks, Andrew. 2019. *Clothing Poverty*. London: ZED.

Carr, Chantel. 2023. Repair and Care: Locating the Work of Climate Crisis. *Dialogues in Human Geography* 13 (2): 221–239.

Cline, Elizabeth. 2013. *Overdressed*. New York: Penguin.

Corbett, Sarah P. 2024. *The Craftivist Collective Handbook*. London: Unbound.

Corwin, Julia Eleanor. 2018. "Nothing is Useless in Nature": Delhi's Repair Economies and Value-Creation in an Electronics "Waste" Sector. *Economy and Space* 50 (1):14–30.

Corwin, Julia, and Vinay Gidwani. 2021. Repair Work as Care: On Maintaining the Planet in the Capitalocene. *Antipode* 57(5):1685-1704.

Crawford, Margaret. 2014. The Garage Sale as Informal Economy and Transformative Urbanism. In The Informal American City, edited by Vinit Mukhija, and Anastasia Loukaitou-Sideris, 21–39. Boston: MIT.

Crewe, Louise, and Nicky Gregson. 1998. Tales of the Unexpected: Exploring Car Boot Sales as Marginal Spaces of Contemporary Consumption. *Transactions of the Institute of British Geographers* 23 (1):39–53.

D'Avella, Matt. 2015. Minimalism. Catalyst. Film.

El-Dorghamy, A. 2009. Youth Viewpoint: For the Love of Egypt. Al-Ahram Weekly. 23–29. Oct.

Fahmi, Wael. 2005. The Impact of Privatization of Solid Waste Management on the Zabbaleen Garbage Contractors in Cairo. *Environment and Urbanization* 17:155–170.

Fahmi, Wael, and Keith Sutton. 2010. Cairo's Contested Garbage: Sustainable Solid Waste Management and the Zabaleen's Right to the City. *Sustainability* 2:1765–1783.

Franck, Karen, and Quentin Stevens, eds. 2007. *Loose Spaces*. London and New York: Routledge.

Fredericks, Rosalind. 2018. *Garbage Citizenship: Vital Infrastructures of Labor in Dakar, Senegal*. Durham: Duke University Press.

Furniss, Jamie. 2014. Alternative Framings of Transnational Waste Flows: Reflections Based on the Egypt-China PET Plastic Flow. *Royal Geographical Society* 47 (1): 24–30.

Graham-Gibson, J. K. 2008. Diverse Economies: Performative Practices for 'Other Worlds'. *Progress in Human Geography* 32 (5): 613–632.

Gregson, Nicky, and Lousie Crewe. 2003. *Second-Hand Cultures*. Oxford: Berg.

Gregson, Nicky, Mike Crang, Julie Botticello, Melania Calestani, and Anna Krzywoszynska. 2016. Doing the 'Dirty Work' of the Green Economy:

Resource Recovery and Migrant Labour in the EU. *European Urban and Regional Studies* 23 (4): 541–555.

Hackney, Fiona, Clare Saunders, Joanie Willett, Katie Hill, and Irene Griffin. 2020. Stitching a Sensibility for Sustainable Clothing: Quiet Activism, Affect and Community Agency. *Journal of Arts & Communities* 10 (1–2): 35–52.

Hansen, Karen Tranberg. 2000. *Salaula. The World of Secondhand Clothing and Zambia*. Chicago: University of Chicago Press.

Harper, Douglas A. 1987. *Working Knowledge*. Chicago: University of Chicago Press.

Haynes, K. E., and S. M. El-Hakim. 1979. Appropriate Technology and Public Policy: The Urban Waste Management System in Cairo. *Geographical Review* 69:101–108.

Herrmann, Gretchen M. 2006. Garage Sales Make Good Neighbors: Building Community Through Neighborhood Sales. *Human Organization* 65 (2): 181–191.

Herrmann, Gretchen M. 2019. Reminiscence and Recompense: Reuse and the Garage Sale. *Worldwide Waste: Journal of Interdisciplinary Studies* 2 (1): 2.

Herrmann, Gretchen M., and Stephen M. Soiffer. 1984. For Fun and Profit: An Analysis of the American Garage Sale. *Urban Life* 12 (4): 397–421.

Hou, Jeffrey, ed. 2010. *Insurgent Spaces*. New York: Routledge.

Hou, Jeffrey. 2014. Making and Supporting Community Gardens as Informal Urban Landscapes. In *The Informal American City*, ed. Vinit Mukhija and Anastasia Loukaitou-Sideris, 79–96. Cambridge: MIT Press.

Houston, Lara. 2019. Mobile Phone Repair Knowledge in Downtown Kampala: Local and Trans-Local Circulations. In *Repair Work Ethnographies Materiality*, ed. Ignaz Strebel et al., 29–160. Singapore: Palgrave Macmillan.

Isenhour, Cindy, and Brieanne Berry. 2020. "Still Good Life": On the Value of Reuse and Distributive Labor in "Depleted" Rural Maine. *Economic Anthropology* 7 (2): 293–308.

Iskander, Mai. 2009. *Garbage Dreams*. Cinema Guild. DVD.

Johnson, Bea. 2013. *The Zero Waste Home*. New York: Scribner.

Johnson, Sarah. 2024. Castoffs to Catwalk: Fashion Show Shines Light on Vast Chile Clothes Dump Visible from Space. *Guardian*, 8 May.

Kamel, Leila I. 1994. *Moqattam Garbage Village*. Cairo: Stallion Graphics.

Kamel, Leila I. 2009. Cairo: A Story of Waste Mis-management. *Al-Ahram Weekly*, 22–28 October.

Kuppinger, Petra. 2021. Sustainability and Small Businesses in Stuttgart, Germany. In *Emergent Spaces*, ed. Petra Kuppinger, 207–226. New York: Palgrave Macmillan.

Kuppinger, Petra. 2023. Incidental Sustainability: Notes from a Thrift Store in Germany. *Critique of Anthropology* 43 (3): 269–288.

Lapolla, Garibaldi. 2009 [1935]. *The Grand Gennaro*. Rutgers University Press.

Le Zotte, Jennifer. 2017. *From Goodwill to Grunge: A History of Secondhand Styles and Alternative Economies.* Chapel Hill: University of North Carolina Press.

Machado, M. A. D., S. O. D. Almeida, L. C. Bollick, and G. Bragagnolo. 2019. Second-Hand Fashion Market: Consumer Role in Circular Economy. *Journal of Fashion Marketing and Management* 23 (3): 382–395.

Mares, Teresa, and Devon Peña. 2010. Urban Agriculture in the Making of Insurgent Space in Los Angeles and Seattle. In *Insurgent Spaces*, ed. Jeffrey Hou, 241–254. New York: Routledge.

Medina, Martin. 2007. *The World's Scavengers.* Lanham: Altamira Press.

Milbourne, Paul. 2010. Growing Places: Community Gardening, Ordinary Creativities and Place-Based Regeneration in a Northern English City. In *Spaces of Vernacular Creativity*, ed. Tim Edensor et al., 141–154. London: Routledge.

Millar, Kathleen. 2018. *Reclaiming the Discarded.* Durham: Duke University Press.

Miller, Benjamin. 2000. *Fat of the Land.* New York: Basic Books.

Minter, Adam. 2019. *Secondhand: Travels in the New Global Garage Sale.* New York: Bloomsbury.

Murphy, Fiona. 2017. Austerity Ireland, the New Thrift Culture and Sustainable Consumption. *Journal of Business Anthropology* 6 (2): 158–174.

Nkatha, Karen. 2023. How Fast Fashion Is Fueling the Fashion Waste Crisis in Africa. Greenpeace, 21 November. https://www.greenpeace.org/africa/en/blog/54589/how-fast-fashion-is-fuelling-the-fashion-waste-crisis-in-africa/. Accessed 2 October 2024.

Pellow, David N. 2004. *Garbage Wars.* Cambridge, MA: MIT Press.

Rashed, D. 2002. Dumping the Zabaleen. *Al-Ahram Weekly*, 11–17 July.

Rathje, William, and Cullen Murphy. 2001. *Rubbish!* Tucson: University of Arizona Press.

Rogers, Heather. 2005. *Gone Tomorrow.* New York: The New Press.

Royte, Elizabeth. 2006. *Garbage Land.* New York: BackBayBooks.

Saraf, Ishani. 2020. "We Too Have Blackened Our Hands": Work, Harm, and Legitimacy in a Delhi Scrap Market. *Anthropology of Work Review* 41 (2): 119–127.

Schmid, Benedikt. 2019. Repair's Diverse Transformative Geographies: Lessons from a Repair Community in Stuttgart. *Ephemera* 19 (2): 229–251.

Schmid, Benedikt. 2020. *Making Transformative Geographies.* Bielefeld: Transcript.

Slackman, Michael. 2009. Cleaning Cairo, but Taking a Livelihood. *New York Times*, 25 May.

Smith, Joe, and Petr Jehlička. 2013. Quiet Sustainability: Fertile Lessons from Europe's Productive Gardeners. *Journal of Rural Studies* 32:148–157.

Sokolovsky, Jay. 2018. Elders, Community Gardens, Civic Ecology, and the Quest for Community. In *Urban Life*, ed. George Gmelch and Petra Kuppinger, 6th ed., 399–412. Long Grove: Waveland Press.

Steward, Shelly. 2020. What Does That Shirt Mean to You? Thrift-Store Consumption as Cultural Capital. *Journal of Consumer Culture* 20 (4): 457–477.

Strasser, Susan. 2000. *Waste and Want*. New York: Holt.

Terry, Beth. 2012. *Plastic Free*. New York: Skyhorse Publishing.

Thomas, Dana. 2019. *Fashionopolis*. New York: Penguin Press.

Tironi, Martin. 2019. Inquiring and Experimenting with Urban Ecology: Pragmatist Lessons from Public Bicycle Repair. In *Repair Work Ethnographies*, ed. Ignaz Strebel et al., 187–219. Singapore: Palgrave Macmillan.

Truitt, Allison. 2012. The Viet Village Urban Farm and the Politics of Neighborhood Viability in Post-Katrina New Orleans. *City & Society* 24 (3): 321–338.

Volpi, Elena. 1996. *The Zabbalin Community of Muqattam. Cairo Papers in Social Science* 19(4).

Williams, D. 2009. Swine-Flu Slaughter Leaves Cairo Without Pigs to Devour Trash. *Bloomberg*. http://bloomberg.com/apps/news?pid=newsarchive&sid=aMODvyH1OBAE. Accessed 24 June 2010.

Zola, Emile. 2012. *The Ladies' Paradise*. Oxford: Oxford University Press.

PART I

Spaces and Practices of Second-Hand Cultures

CHAPTER 2

Reuse and ReUse: Informal and Institutional Transfer of Used Goods

Gretchen Herrmann

Reuse of pre-owned goods is increasingly recognized as an important practice to reduce consumer waste and benefit the planet through developing a circular economy. There are numerous pathways to pursue reuse, including on-line groups such as Craigslist, trashnothing, freecycling, eBay, and Facebook Marketplace. Face-to-face venues for selling second-hand goods include flea markets, garage sales, boot sales, and thrift stores. Each method for selling cast-off used goods has a particular mode of operation and resulting social relations. This chapter examines and compares the styles and social relations enacted in a not-for-profit brick and mortar store dedicated to "recycling,"[1] called the Finger Lakes ReUse Center, and garage sale exchanges. Located in Ithaca, NY, the ReUse Center plays a significant role in providing future use for large quantities of still functional second-hand items.

The ReUse Center, or simply ReUse, as it is locally known, defines its mission as promoting the continued use of second-hand goods to benefit the environment, economy, and local community. This nonprofit runs

G. Herrmann (✉)
Ithaca, NY, USA
e-mail: Gretchen.Herrmann@cortland.edu

P. Kuppinger (ed.), *Second-Hand Cultures and Economies of Reuse, Repair, Sharing, and Care*,
https://doi.org/10.1007/978-3-031-99875-1_2

multiple facilities where workers sort, clean, repair, and sell still useful items like appliances, furniture, clothing, electronics, books, housewares, and much more. The same benefits to the environment and the economy and the creation of community have been documented for garage sales, although they emerge informally there (e.g. Herrmann 1997, 2019, 2023) as the result of house cleaning and trying to make a little money. In this chapter, I examine how both these forms of exchange, the ReUse Center as a brick and mortar permanent institution and garage sales as temporary informal activities, function to enhance the environment through promoting reuse, while developing the local economy, and creating community. Concretely, I focus on the role and future of the things that change hands in each venue. As Adam Minter observes: "Reuse and resale give consumers a rare, tangible connection to the afterlife of their stuff" (2019, xiv).

The Finger Lakes ReUse Center

Thrift stores specializing in second-hand trade have experienced considerable growth in recent decades (e.g. Minter 2019). While many thrift stores, often with a religious affiliation, continue their historic function of generating money and providing goods to assist those in need, the ReUse Center also creates jobs, generates considerable revenue, and fosters community programs. In a college town filled with transients, the Center performs a vital function providing inexpensive second-hand goods for this population and others in need. The welcoming message on the ReUse Center's webpage notes:

> Welcome to Finger Lakes ReUse. We are working toward a just, resilient and waste-free world that values people and resources. Join us in reducing waste sent to landfills, and giving quality materials and people new chances, all while making a local impact. (https://ithacareuse.org)

A local journalist observes, "ReUse is something of a unique organization. Every good-sized town has its Salvation Army or Goodwill. But ReUse is more than that" (Crandall 2019). For example, a ReUse crew deconstructs buildings and sells salvaged parts. ReUse also provides assistance for the technologically challenged and job training programs. ReSET seeks "to help get people who have had difficulties in entering or staying in the workforce skills they can use to obtain gainful employment"

(Crandall 2019). There has been a significant job or college placement rate for those completing the training and apprenticeship program" (Crandall 2019).

Founded in 2005 with a small grant, a number of ecologically minded citizens began planning a ReUse Center, and the doors of ReUse opened in 2008. Diane Cohen has served as the Executive Director since then, and she has maintained a stellar reputation in the field. In January 2022, she received a lifetime achievement award for recycling from the New York State Association for Reduction, Reuse, and Recycling, which adds to previous honors such as a 2017 eTown eChievement Award (Wickham 2022). Today, the ReUse Center is a $3 million organization operating two sites (Cohen 2023). The downtown site, Ithaca ReUse, sells a full range of household goods, furniture, and even some building parts, such as used doors, while the Triphammer Road ReUse MegaCenter on East Hill sells clothing, furniture, electronics, books, and CDs. The ReUse Center also started a facility in nearby Cortland, which has since become independent.

The organization embraces reuse as a way to develop the local community and economy, but the ecological reuse function of keeping still useful goods out of the landfill, is at the core of their mission. They estimate that about $2000 to $3000 revenue is generated by ReUse for each ton recovered, in contrast to the price of $96 per ton for tipping fees at a transfer station (Elliott and Cohen 2022). ReUse workers labor to keep the shelves stocked and densely packed because they have determined that for every $100 in merchandise on the floor, they generate $70 of sales (Crandall 2019).

The ReUse Center has experienced a 20 percent growth in business each year for the past 10 years along with other praiseworthy statistics. For example, about 80 employees receive living wages and insurance along with 10 youth workers and apprentices. In 2022, the Center received 45,000 drop-off donations and approximately 750,000 items were returned to local use (Cohen 2023). Currently, the Reuse Center is about 80 percent self-sufficient and strives to be completely self-sufficient. Cohen believes they have done so well over the years because of their strategy of "convenience, affordability and visibility" (Crandall 2019). An economic impact study of Finger Lakes ReUse found sales revenue generated at $8.67million from 2018 to 2022 (Mikec and Augustine-Marcell 2023).

The ReUse Center provides a vast variety and, very importantly, high degree of reliability in goods sold not found in short-lived, informal garage sales. ReUse workers examine electronics and other items to determine that they work and provide a short "guarantee" for such goods along with a sticker attesting to their functionality. The permanent storefront provides a place to return goods over a longer time.

The ReUse Center is both constrained and empowered by forces and regulations that apply to all businesses. The ReUse Center transforms reusable goods into paid jobs on a large scale and sizable volume. It creates significant economic value by revaluing goods that are otherwise headed for landfill oblivion. They are part of the distributive economy wherein labor involved in shopping, selling, and reclaiming goods at these localized sites is valorized and contributes to the formal economy (see Herrmann 2019; Isenhour and Berry 2020). Profits made off "trash" are reinserted into the local economy.

The ReUse Center builds community in multiple ways. It provides community meeting areas and focuses on meaningful activities, such as a repair clinic. It provides goods at low prices for those in need. It generates living wage employment, teaches work skills to various constituencies, such as formerly incarcerated individuals and people with disabilities. Diane Cohen notes, "moving someone off of public assistance and into living wage work has tremendous social value" (Crandall 2019). The Center offers community programming for diverse residents. The ReUse Center and allied academic and civic sponsors recently hosted a "New York State ReUse Summit," drawing attendees from all over the state and beyond. The Finger Lakes ReUse model has generated so much interest beyond its local context that the Center developed a web-based template for groups that want to establish their own ReUse centers. The ReUse Center has even set up a taco truck, run by a Mexican American, in the Center's downtown parking lot. They hold periodic tent sales, such as the one before Labor Day to catch student shoppers and others setting up new households. They run sales when they are overstocked, like a recent 35 percent off books event at the Mega Center. These activities are purposefully undertaken to enhance the quality of support, local prosperity, and environmentally conscious behaviors.

Thrift stores have been increasingly inundated with unsellable items, such as stained clothing, broken ceramics, or damaged electronics. Referred to as "donation dumping" (Minter 2019; Berry 2022), such items must be discarded by the thrift workers, tying up their labor, overwhelming sorting areas, and producing additional costs. One way the Reuse Center minimizes losses is by screening donations and turning away unusable or irreparable goods. This allows the Center to staff the screening activities rather than examining mounds of questionable goods, many of which would need to be landfilled (Berry 2022; Minter 2019). Screening donations frees up space needed for sorting and display and reduces stress on personnel (Berry 2022; Lamb 2021).

As part of the professionalization of thrift stores in the UK and elsewhere, they usually do not allow bargaining to lower prices (Fitton 2022). However, many workers do in fact bargain with customers for a variety of reasons, most commonly because of damage or irregularities in the items, such as rips, stains, or holes. The workers at the ReUse Center do not usually bargain. Instead, the store uses color-coded tags that indicate prices which staff lower 25 percent from the original price every 60 days. I once witnessed a futile attempt by a customer to bargain at the ReUse Mega-Center: a man tried to remove an unwanted picture frame from an art print for a reduced price and was rebuffed by a clerk, who reminded the customer that the regularly scheduled discount of 25 percent would take effect in two days. In the ReUse Center, customers are less able to acquire used goods by way of negotiating their prices and adding that to the item's social history, as they can at garage sales.

Garage Sales

A garage sale is defined here as the temporary public sale of used goods, primarily household items and clothing, from, in, and around a private residence. Garage sales are held in most American communities. They are organized and attended by diverse constituencies, although middle and working-class participants predominate. Garage sales can bring shoppers and sellers from different backgrounds into informal contact. The salient factor is that, since sales transpire in and around the home (i.e., that public commerce and proprietary hospitality are juxtaposed), participants are selling (more or less) their own goods. This forms the stage for broad social interactions and relations (Herrmann 1997), ranging from the impersonal and commercial to the more intimate and gift like. Sellers participate with a wide range of styles and motivations, from periodic housecleaning and socializing to generating cash and profit making. Shoppers' motivations include recreation, recycling, and bargain hunting. Most participants do not derive major financial benefits from garage sales, but dealers and those in financial need may be especially profit oriented. Public and private are brought together in an unusual synthesis in garage sales (Herrmann 1997). Women, the primary garage sale participants, combine domestic concerns with public commerce to play a key role in creating this mixed and multilayered form of exchange. As I have demonstrated in previous publications, garage sales generate community (Herrmann 2006), economic benefits (e.g., Herrmann 1997), and promote ecologically minded reuse (Herrmann 2019). They do so as a result of ridding homes of unwanted or no longer needed objects.

Many garage sale hosts offer their items for sale to prevent goods from ending up in the landfill. Some municipalities, such as Sunnyvale CA, hold community-wide garage sales expressly to reduce tonnage entering the landfill (Herrmann 2019). Australians, too, have enthusiastically adopted garage sales to sell "pre-loved" items in a mass party-like atmosphere like the "Garage Sale Trail." Entire states may hold sales at the same time, often before the Christmas giving season. One local council posted that in 2023 more than 400,000 people shopped at over 12,000 sales held during the Garage Sale Trail (NSW epa 2023). They estimated five million pre-loved items had been for sale and made a considerable ecological impact. The sponsor of this event was the NSW Waste and Sustainable Materials Initiative, funded from the waste levy. Its primary mission was to lighten the landfill load.

The Future Life of the Discarded?

Anthropologist Fabio Gygi distinguishes between relation friendly and nonrelation friendly styles of ridding household excess in the Japanese context. He highlights passing along of affect and an ability to release hold of objects and promote relationships with buyers, sellers, and the objects themselves in the personalized garage sale, especially when sellers envision a future life in a new "good" home for an object (Gygi 2022, 13; Herrmann 2015). Gygi views garage sale goods as "anti-alienating" objects in that they were originally plastic entombed commodities, but now carry emotions, histories, and wear and tear, and are hence relation friendly. He sees the social process of transferring them in the garage sale as a relation friendly practice. This contrasts with the regime of the Japanese declutter guru Marie Kondo (2014), who focuses on the elimination of unwanted items, something she finds best selected in solitude rather than in a social setting. Gygi views this process as relation unfriendly since there is no regard for the history or future of these goods. To Gygi these activities also rebalance cultural orientations, like in garage sales, where objects are personalized and affect imbued, whereas, in the Kondo method, belongings are largely removed from human emotions and attachment. The process of removing items is not interactive in her declutter protocol and does not generate nascent human relations the way reselling goods can in the more personal garage sale exchange.

Items rejected using the Kondo method, despite a token thank-you for their service, are relegated to the category of waste. Joshua Reno asserts that waste is essentially the by-product of organization (2016) and the Kondo method creates winners and losers, retained objects and trash bin detritus. Second-hand items sold at in-person venues were once considered waste (or being useless for their previous owner), because they were rejected from a household, but this rejection is ameliorated through the personalized transfer of the affect-laden goods and their eventual resale. Both the more impersonal context of a thrift store like ReUse and more personal or sociable garage sales provide a meaningful "afterlife" for discarded items. Items decluttered via a Kondo-like discarding activity might end up in a landfill, as the main focus here is their uselessness or superfluous existence in their current environment.

Narrative Sociality of Used Goods

Scholars have documented the rich narrative sociality that is part of second-hand exchanges (Appelgren and Bohlin 2015, 151–4). Buyers and sellers relish stories of objects' origins and their histories. Swedish sellers may provide the history of jeans for sale and second-hand purchases may always be known as *second-hand*. One of my garage sale interlocutors embraced this storytelling spirit visible in the various modes of second-hand exchange like this:

> Everything has a *history* and I enjoy passing on the history to someone who might use it. And when I buy something, I'd be interested in knowing who has had that. And it becomes a social-historical kind of exchange in the best garage sales, because you want to know, because it's coming into your possession, so you want to know whose possession it was before. And why it was precious to somebody else at one time. So everything almost can have a story to it and I like stories. So, I envision that with someone it will certainly find a home someplace. A nice thing that someone will be very happy to receive and I will tell them the history.

These stories add provenance to the items and help "singularize" them (Kopytoff 1986). They are found especially in venues such as flea markets and garage sales where there is the likelihood of exchange with the seller of the objects who is familiar with their history and can relate it to buyers. Such stories even surface in virtual sales like on eBay. "Circulating objects pass through dense social contexts where layers of events, memories, and meanings motivate people to interact, narrate stories, and form relationships" (Appelgren and Bohlin 2015, 152). Many memories and much affect may be embedded in some objects that Sasha Newell refers to them as "hard drives" (2014, 198). Octave Debary observes that face-to-face sales in France are "where personal stories change hands [and] where new memories are cobbled together from old objects" (2015, 123). One interlocutor noted that, "Whenever people tell a story, the object has more of an essence, more of a meaning. They're attaching something that's not just an object. It's an object which is connected to someone's life." Further, to the extent participants grant animism or agency to objects, perhaps by selecting a new owner, the objects would similarly enliven both a ReUse type second-hand store and neighborhood garage sales.

The more anonymous nature of items in thrift stores might make customers fantasize about the origins and histories of old items. They might imagine fictive stories about second-hand finds such as retro clothing (e.g. Gregson and Crewe 2003, 5), since second-hand shops often spark the imagination. Melanie Lovatt (2015) finds that the imagined future of possessions of a deceased family member's belongings through a charity thrift shop allows family members to let go of items they were ambivalent about keeping. Such donations are motivated by the notion that the items might be useful to others in the future. While the donor will not know the future owner of clothes and other items, they can still feel good about these items and that they will hopefully find a "loving" new home. This hope prompts them to donate these things to a thrift store. Similarly, professional organizers may encourage clients to imagine stories about the future of possessions because that makes it easier for clients to let go and donate the items, with aspirational stories attached (Kilroy-Marac 2016, 449).

Adam Minter confesses he had to have faith that the goods donated from his mother's estate would be used in some way, that they would be valued and not landfilled or incinerated (2019, xv.) Roster (2001) uses the term providing "safe passage" to describe the stories that surround the journeys of second-hand items. I prefer to describe them as adding provenance in the transfer of second-hand goods. Provenance and legacy tie the goods to the family life where they come from and through which they extend their biographical pathway often by exchange between "kindred spirits" (see Herrmann 2023) with its overlay of family and kinship.

Goods and People in Thrift Stores and Garage Sales

Small and large thrift stores and garage sales exist in the same larger universe of discarded but still useful second-hand goods. However, they occupy different, yet ultimately complementary places in this landscape. Because of their size and permanent nature, thrift stores, like the ReUse Center, can offer customers predictably and reliably a very large assortment of second-hand goods. They are there regardless of season or weather. They spend much time screening, sorting, checking, and possibly repairing donated items. They even might give very short-term store "guarantees" for some items. Because of their permanent nature, customers might return items or at least be able to complain about things. Garage sales, because of the ephemeral nature, provide none of this. Sales are permanent and there is not really much of a chance to go back and complain (unless, of course one is a next-door neighbor or knows the seller well). Garage sales mostly also offer things at even lower prices than thrift stores, as they have no overhead costs at all.

Places like the ReUse Center are constrained and empowered by legal regulations that frame their work. Having a permanent building and employing workers requires more rules and is more costly than setting up a few tables and clothes racks in one's driveway. There are fire regulations or workers' benefits to be considered. None of this applies to garage sales and makes their operation easier, but also short-lived. The ReUse Center transforms reusable goods into paid jobs, which is a valuable community service. Garage sales have no such benefits.

Both thrift stores and garage sales turn "trash" into value as they present what might have been destined for the landfill to those who need and want to provide a second life for these things. They are part of a distributive or local circular economy. The labor of reclaiming and even of shopping in these contexts is valorized and contributes to the formal economy (see Herrmann 2019; Isenhour and Berry 2020). Sales in both contexts take "discards" and create (small) profits that stay in the local economy. Garage sales provide face-to-face opportunities to exchange stories and histories and thus are richer sources of narrative sociality than the ReUse Center. Yet even ReUse, like most thrift stores, occasionally allows customers to brush up against the history of some goods. Objects for sale in the ReUse Center are largely removed from their contexts of previous use and memories, and do not generate the same degree of interaction and transfer of affect between shopper and seller as do garage sales. Nevertheless, goods may still carry some transmissible emotion and bear the traces of previous use (e.g., names written in children's clothes), as well as the shared value of reuse and civic pride. Shoppers and donors alike experience some generalized good feeling for supporting an institution that does good in and for the community. Many community members routinely shop there, and the store facilities are always busy. Much of the affect transmitted in the ReUse Center is collective in nature, an amalgam of shared community values, contrasting with the personalized memory and emotion afforded by garage sales.

While buyers in the garage sale and ReUse Center generally like what they purchase and enjoy the memories and emotions transmitted by sellers, they may not feel a commitment to keep the objects "forever." Anna Bohlin (2019) notes that in Sweden shoppers' commitment to second-hand-ness may be more important than their long-term enticement by a particular object. They can enjoy an item for a while then let it go back into the second-hand stream without guilt, as keeping things out of the landfill and circulating is their paramount motivation.

Conclusion

Although both the ReUse Center and garage sales trade in similar second-hand goods, they provide somewhat different experiences of exchange. Overall, garage sale exchanges are more family and home related, while trade in the ReUse Center is more public and community oriented. In the garage sale community ties develop through the transfer of stories and affect with the goods exchanged, positive in-person interactions with neighbors and friends, and frequent "giving" to those who appear in need. In contrast, social community development in the ReUse Center is a conscious social creation. There are formal programs for youth, diversity, and inclusion, those in need, and a collective for those engaged in repair. Both venues promote economic, environmental, and community benefit, but they ultimately have differing forms of connectedness as part of the exchanges involved. Garage sale trade is prosocial and involved with relationships to the object, examining both its past and future use, whereas the ReUse Center is an institutional setting in which most unique affective attributes have been separated from the goods transferred. Exchange here is more oriented to the functionality and design of things, rather than their past or future use as such. Yet some affect from individuals may attach to such objects as a leather belt, or a wooden bowl, and the imaginings of positive use of items has been shown to spur donations.

A core difference between institutions like ReUse and garage sales are issues of volume, permanence, and predictability. ReUse is there, rain and shine, summer and winter, is always well stocked and ready to help. If one needs a toaster or a pair of jeans, one is very likely to find these objects, and if not this week, then next week or the week after. Garage sales pop up and provide an unpredictable and eclectic mix of things in an often party-like atmosphere. They provide stuff and fun but are not reliable sources of concrete items. They are entertaining and individualistic. ReUse has a larger mission and goals that are more comprehensive. Its offerings bespeak collective sentiments such as pride in community, value of the environment, avoidance of rampant consumerism, and social justice. In the end, the Finger Lakes ReUse Center aims at a much larger scale ecologically minded avoidance of unnecessary landfill dumping, and allows all involved to feel good about what can be salvaged.

Note

1. Minter (2019, xiv) points out that most people refer to "reuse" as "recycling," a usage I also found among garage sale participants (e.g. Herrmann 2019).

References

Appelgren, Stephan, and Anna Bohlin. 2015. Growing in Motion: The Circulation of Used Things on Second-Hand Markets. *Culture Unbound* 7:143–168.

Berry, Brieanne. 2022. Glut: Affective Labor and the Burden of Abundance in Second-Hand Economies. *Anthropology of Work Review* 43 (1): 26–37.

Bohlin, Anna. 2019. "It Will Keep Circulating": Loving and Letting Go of Things in Swedish Second-Hand Markets. *Worldwide Waste: Journal of Interdisciplinary Studies* 2 (1): 1–11.

Cohen, Diane. 2023. Community ReUse Center Model. https://cdrpc.org/wp-content/up. Accessed 1 April 2024.

Crandall, Brian. 2019. At Finger Lakes ReUse, Every Day is Thanks (for) Giving. *Ithaca Voice*, November 29.

Debary, Octave. 2015. Yard Sales: A Book and an Exhibition from Selling Off Objects to Redeeming Memory. *Culture Unbound* 7:123–142.

Elliott, Robin, and Diane Cohen. 2022. Reuse Summit Generates Interest in Finger Lakes ReUse. *Tompkins Weekly*, August 24–30:15–17.

Fitton, Triona. 2022. Pricing Up & Haggling Down: Value Negotiations in the UK Charity Shop. *JOMEC Journal* 17:55–78.

Gregson, Nicky, and Louise Crewe. 2003. *Second-Hand Cultures*. New York: Berg.

Gygi, Fabio. 2022. Falling in and Out of Love with Stuff: Affective Affordance and Horizontal Transcendence in Styles of Decluttering in Japan. *Japanese Studies* 42 (2): 195–212.

Herrmann, Gretchen M. 1997. Gift or Commodity: What Changes Hands in the US Garage Sale? *American Ethnologist* 24 (4): 910–930.

Herrmann, Gretchen M. 2006. Garage Sales Make Good Neighbors; Building Community through neighborhood Sales. *Human Organization* 65 (2): 181–191.

Herrmann, Gretchen M. 2015. Valuing Affect: The Centrality of Emotion, Memory and Identity in Garage Sale Exchange. *Anthropology of Consciousness* 26 (2): 170–181.

Herrmann, Gretchen M. 2019. Reminiscence and Recompense: Reuse and the Garage Sale. *Worldwide Waste: Journal of Interdisciplinary Studies* 2 (1): 1–10.

Herrmann, Gretchen M. 2023. The American Garage Sale: Liberating Space and Creating Kin. In *Of Hoarding and Housekeeping*, ed. Sasha Newell. London: Berghahn.

Isenhour, Cindy, and Brieanne Berry. 2020. "Still Good Life": On the Value of Reuse and Distributive Labor in "Depleted" Rural Maine. *Economic Anthropology* 7 (2): 293–308.

Kilroy-Marac, Katie. 2016. A Magical Representation of the Modern: Professional Organizers and Thingly Care in Contemporary North America. *Cultural Anthropology* 31 (3): 438–457.

Kondo, Marie. 2014. *Life Changing Method of Tidying Up*. Berkeley, CA: Ten Speed Press.

Kopytoff, Igor. 1986. The Cultural Biography of Things: Commoditization as Process. In *The Social Life of Things*, ed. Arjun Appadurai, 64–91. Cambridge: Cambridge University.

Lamb, Anna. 2021. ReUse Center Pausing Donations Due to High Volume Already Received. *The Ithaca Voice*, June 8.

Lovatt, Melanie. 2015. Charity Shops and the Imagined Future of Objects. *Culture Unbound* 7:13–23.

Mikec, Nika, and Wyeth Augustine-Marcell. 2023. Quantifying the Economic Impact of Finger Lakes ReUse in Tompkins County, New York. *Finger Lakes ReUse*. https://ithacareuse.org. Accessed 1 December 2024.

Minter, Adam. 2019. *Second-Hand: Travels in the New Global Garage Sale*. New York: Bloomsbury Publishing.

Newell, Sasha. 2014. The Matter of the Unfetish: Hoarding and Spirit Possession. *HAU* 4 (3): 185–213.

NSW epa. 2023. Garage Sale Trail. nsw epa 2023. Accessed 1 March 2024.

Reno, Joshua. 2016. *Waste Away: Working and Living with a North American Landfill*. Oakland, CA: University of California Press.

Roster, Catherine A. 2001. Letting Go: The Process and Meaning of Dispossession in the Lives of Consumers. *Advances in Consumer Research* 28:425–430.

Wickham, Jessica. 2022. Diane Cohen Receives State Award for Work at ReUse. *Tompkins Weekly*, January 26.

CHAPTER 3

Social Capital in Secondhand Markets

Brieanne Berry, Cindy Isenhour, Andrew Crawley, and Sujan Chakraborty

Introduction

It's an overcast summer day in Mill Town,[1] a rural community in Maine, and Mrs. Morgan's yard sale is deserted. Customers who stop in her driveway are greeted with the well-rehearsed script: "Hi there! How are ya? It's

B. Berry (✉)
Department of Environment & Sustainability, Ursinus College, Collegeville, PA, USA
e-mail: bberry@ursinus.edu

C. Isenhour
Department of Anthropology/Climate Change Institute, University of Maine, Orono, ME, USA
e-mail: Cynthia.isenhour@maine.edu

A. Crawley
School of Economics, University of Maine, Orono, ME, USA
e-mail: andrew.crawley@maine.edu

S. Chakraborty
Department of Economics, Indiana University—Indianapolis, Indianapolis, IN, USA
e-mail: chakras@iu.edu

P. Kuppinger (ed.), *Second-Hand Cultures and Economies of Reuse, Repair, Sharing, and Care*,
https://doi.org/10.1007/978-3-031-99875-1_3

mostly kids' clothes. Little girls' clothes. So if that's what you're looking for, you may have hit the jackpot!" While one or two shoppers stay to look at her tables full of neatly folded children's clothes, many depart without browsing, let alone purchasing any items. Resigned to the fact that many of her items will go unsold, Mrs. Morgan explains that "there's a lot of low-income families [in the area]. I think it's more than half that qualify for some sort of aid from the schools, so there's definitely families that could benefit from [this stuff], it's just figuring out how to get it into their hands [that] is the challenge" (Interview, 8/17/19). Motivated to sell her children's outgrown clothes in part by a desire to help her community, Mrs. Morgan is similar to many of the reuse participants we heard from over the course of our five-year research project on secondhand economies in Maine. Her struggle to connect her used goods with people in need illuminates important questions about how social networks shape access to used goods as well as how goods themselves can shape and build social relations.

Reuse is commonly understood as a sustainability strategy with the potential to reduce waste and environmental degradation by using products longer and reducing the need to manufacture new goods. It is a key component of the conceptual "circular economy" which, in contrast to linear take-make-waste systems, is intended to be restorative and regenerative, in part, by slowing resource loops by keeping existing products in use longer. Importantly, alongside its environmental benefits, reuse is often associated with social benefits, as well. Some proponents argue that collaborative forms of reuse can improve social capital by re-embedding exchange in social and ethical logics (Barnett et al. 2005; Carrier and Luetchford 2012), rebuilding social ties weakened by geographic mobility and individualized lifestyles (Jones et al. 2015; Jordan 2015), increasing psychological and emotional well-being (Grobsmith 1981), and contributing to resilience by creating localized ties of reciprocal support in communities without consistent access to retail and employment markets (Isenhour 2003). Indeed, for the Ellen MacArthur Foundation, one of the leading proponents of circular economies internationally, building social capital (alongside economic and natural capital) is a key component of the definition of the circular model (EMF 2022).

Broadly defined as the idea that social networks have value (Putnam 2000b) and can help individuals and communities achieve collective goals (Coleman 1988), social capital is a contested (Durrenberger 2002), although much-used concept in academic and policy literature. The concept is making a particular resurgence lately in relation to our ability to adapt to crises like COVID-19 (Pitas and Ehmer 2020; Borgonovi and Andrieu 2020), natural disasters (Barnett et al. 2005) and climate change (Hamilton and Lubell 2019; Benevolenza and DeRigne 2019). Similarly, resource pressures, pollution, and the climate impacts of linear economic systems have recently driven academic interest in reuse markets. However, the connections between reuse and social capital are based on a kind of received wisdom—a sense that secondhand exchanges naturally build upon and lead to the further development of social networks, trust, and reciprocity (Botsman and Rogers 2010). There is also speculation that strong systems for reuse can improve community resilience when disasters or shocks disrupt supply chains and markets (Aytac et al. 2016).

Drawing on a statewide household survey of reuse market participants ($n = 605$) and ethnographic engagement including participant observation and in-depth interviews with buyers, sellers, donors, and recipients of used goods ($n = 54$), this paper details our attempts to explore these claims and to understand social capital within the context of Maine's vibrant and varied reuse economy. The picture of social capital that emerged from our work is complicated. Rather than a uniformly available collective resource, we found that social capital was patchy and uneven in Maine's reuse markets—benefitting some and excluding others, often based on differences in race and class. So while our research contributes to debates about the potential for reuse markets to facilitate transitions toward more sustainable economic forms, it also provides theoretical and practical insight into social capital with important implications for how reuse policies and projects might be designed to ensure just transitions toward more circular economies.

In what follows we explore the epistemological roots of social capital as a concept before analyzing whether and how social capital is present within secondhand economies in Maine. Next, we consider how our qualitative and quantitative measures succeed at illustrating the nuances and complexity of the relationship between social capital and participation in various forms of reuse. In the end, we argue that if reuse is to be encouraged as a key strategy for transitioning toward more sustainable, circular economies, social benefits are certainly possible, but should not be assumed. However, with intentional design to address issues associated with equity and a just distribution of benefits, reuse can certainly contribute to social well-being.

Literature Review: Capitals

Social capital inhabits a deeply heterogeneous literature (see, e.g., Flora et al. 2015; Bourdieu 2004 [1986]; Coleman 1988; Portes 1998; Adger 2003; Fine 1999; Scrivens and Smith 2013) where its uses, definitions, and measures vary significantly. Often used as a gloss for civic engagement or positive sociality, social capital has complex epistemological roots that are important to acknowledge if the concept is to be used with any nuance.

Scholars have used or alluded to social capital since the early 1900s (e.g., Hanifan 1916) but its epistemological pathway has branched, twisted, and circled, making any concise definition of the concept challenging. In the very broadest sense, social capital can be defined as the idea that "social networks have value" (Putnam 2000b, 11) such that those with more social networks of support have more social capital. We can roughly divide social capital theorists into two schools: the "public good" scholars (Wall et al. 1998), including Coleman (1988) and Putnam (1995a, b, 2000b), who conceive of social capital as an unalloyed good that should be rebuilt in the face of its perceived decline in the United States; and the critical theorists (e.g., Bourdieu 2004 [1986]; Lin 1999; Portes 1998) who see social capital as (re)productive of inequality and exclusion.

For the public good scholars, social capital is an asset that is linked to improved educational outcomes, safer streets, entrepreneurial success, innovation, and national-level economic prosperity (Coleman 1988; Putnam 2000b; Malecki 2012). In this view, "social capital is productive, making possible the achievement of certain ends that in its absence would not be possible" (Coleman 1988, S98). Putnam (2000b) famously documented what he framed as the decline of social capital in the United States, marked in part by fewer people participating in bowling leagues and more people "bowling alone." For Putnam, declining social capital equates to less trust and cooperation within society. This public good approach has been heavily criticized for neglecting issues of class, power, and inequality (Fine 1999; Naughton 2014), being apolitical (Schneider 2008), and for ignoring the ways in which social capital can constrain human behavior in addition to affording opportunities (Portes 1998).

By contrast, critical social capital theorists emphasize the ways in which inequality and exclusion are linked to social capital accumulation. For Bourdieu, social capital is "the aggregate of the actual or potential resources which are linked to [...] membership in a group—which provides each of its members with the backing of the collectively-owned capital, a 'credential' which entitles them to credit" (2004 [1986], 21). This network-based capital is based on the idea of inclusion in social networks but is—conversely—inherently exclusive. At the same time that it provides access to resources for some, it necessarily keeps others out (Schneider 2008). The very closed structures that can support collective action and promote safe communities can suppress individual choices and enforce homogeneity (Portes 1998) by excluding those considered "outsiders" based on a wide array of criteria (e.g., geography, race, ethnicity, gender identity, or income). Other scholars note that the promotion of social capital can support a neoliberal agenda of small government, reduced services, and self-sufficient communities (Naughton 2014). These critical scholars highlight the uneven accumulation and impacts of social capital for individuals, groups, and communities.

Part of the difficulty in integrating critical and public goods approaches is that they are often working toward different aims. For those who see social capital as a public good, including some scholars and many policymakers (e.g., Putnam 2000b; Coleman 1988; Prewitt et al. 2014; Sawhill 2020), the emphasis is on how to measure social capital, for its (positive) outcomes are often assumed. These texts rely largely on quantitative methods to assess the presence or quantity of social capital. Many of the critical theorists who follow Bourdieu take issue with the methods and implicit assumptions of the public goods scholars (e.g., Fine 1999; Narotzky 2007; Durrenberger 2002), yet few offer empirical research that proposes an alternative understanding of social capital. As such, there is a dearth of empirical research that explores the concept of social capital in all of its complexity—that is, using both qualitative and quantitative methods. To take on such a project requires situated knowledge that can explore whether social capital indicators are present *and* what effect they have on people's lived experiences.

Social capital has been operationalized for quantitative research in terms of number of associational memberships (Putnam 2000b; Warde and Tampubolon 2002), number of friendships (Warde and Tampubolon 2002), volunteerism rates (Putnam 2000b), newspaper readership (Putnam 2000b), voter turnout (Putnam 2000b), census participation (Rupasingha et al. 2006), number of bowling establishments (Rupasingha et al. 2006), number of golf courses (Rupasingha et al. 2006), and exchanges of money (Hofferth and Iceland 1998). These variables lend themselves to quantitative analysis and offer a very narrow vision of social capital—one that could easily capture the practices of affluent, white people and communities, while neglecting others'. Indeed, many critical theorists argue that such representations of social capital neglect power, inequality, and class (Schneider 2006), and that a qualitative approach is critical to reintroducing these issues into the theoretical framework (Schneider 2006; Svendsen 2006; Wall et al. 1998; Woolcock 2001).

In what follows, we draw on quantitative and qualitative methods to explore the presence and impact of social capital in reuse markets—as a means to evaluate common claims about the social benefits of reuse markets. We integrate public goods scholarship and the critical theorists' work, exploring the positive potential of social capital that exists alongside its ability to perpetuate inequality within the rural reuse economies where our ethnographic work was based. We break social capital down into four components commonly found in the social capital literature: personal relationships (Apinunmahakul and Devlin 2008; Flora et al. 2015; Leahy and Anderson 2010; Scrivens and Smith 2013), social network support (Demetry et al. 2015; Duncan and Blackwell 2014; Flora et al. 2015; Scrivens and Smith 2013), civic engagement (Aldrich and Meyer 2014; Putnam 2000b; Rupasingha et al. 2006; Scrivens and Smith 2013), and trust and cooperative norms (Adger 2003; Aldrich and Meyer 2014; Coleman 1988; Gallent 2014; Scrivens and Smith 2013). These four components allow us to explore social capital as a combination of interrelated parts with differentiated effects.

Methods: Things in Motion

Our work unfolded across scales, from in-depth ethnographic research within two rural Maine communities to a statewide household survey of reuse participation. This mixed methodology allowed us to measure relationships between reuse and social capital using quantitative methods, as well as to understand the highly contextual relationships between reuse, social capital, power, and privilege using qualitative methods. These methods are complementary rather than comparable—they serve to help us think about the process of measuring and understanding social capital in the context of secondhand economies.

Quantitative Methods: Statewide Household Survey of Reuse in Maine

In order to better understand how and why people participate in reuse markets, as well as how those diverse practices might be connected to social capital, we conducted a statewide survey of Maine households. In addition to gathering demographic and household information, the survey asked participants to self-report the types of exchanges they had engaged in (buying, gifting, selling, bartering, lending) the methods utilized (personal gifts, online sales, brick and mortar purchases, etc.) and the types of good exchanged over the past year. Social capital indicators in the survey included measures of trust (Grootaert and van Bastelaer 2002; Putnam 2000a, b), expectations of social network support (Grootaert and van Bastelaer 2002; Scrivens and Smith 2013), number of close friendships (people you feel at ease with, can talk to about private matters, or call on for help) (Putnam 2000a; Scrivens and Smith 2013), and volunteerism (number of hours volunteered per month on average, organizations volunteered for in the past year) (Hofferth and Iceland 1998; Scrivens and Smith 2013; Putnam 2000a).

The survey was distributed via mail, based on a representative sample of 4000 Maine households with verified postal addresses purchased for the project. We distributed the survey in two separate mailing rounds. The first round of 4000 surveys was mailed in June, 2018. The second round of 3500 was mailed in October, 2018. Participants were invited to complete the paper survey they received in the mail, or to fill out the survey online using a unique participant ID. We received a total of 605 responses, for a 12.4% response rate. An undergraduate research assistant was trained to enter data and a second member of our team double entered 10% of the responses to check for data accuracy.

Survey responses were not representative of Maine's population, with differences around annual household income (see Fig. 3.1), age (see Fig. 3.2), and educational attainment (see Fig. 3.3). In general, our survey population was slightly higher-income, older, and more educated than the average adult in the state of Maine. Notwithstanding these limitations, the sample provides a first attempt to quantitatively assess relationships between participation in reuse markets and measures of social capital.

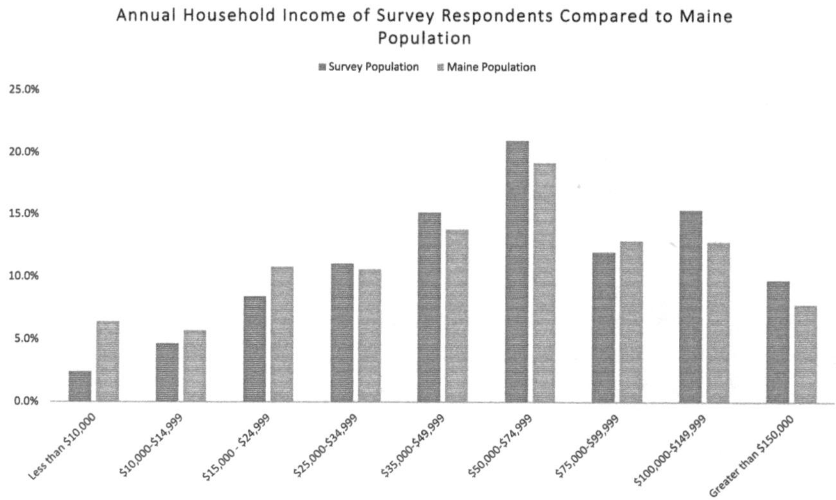

Fig. 3.1 Income of survey respondents compared to Maine population

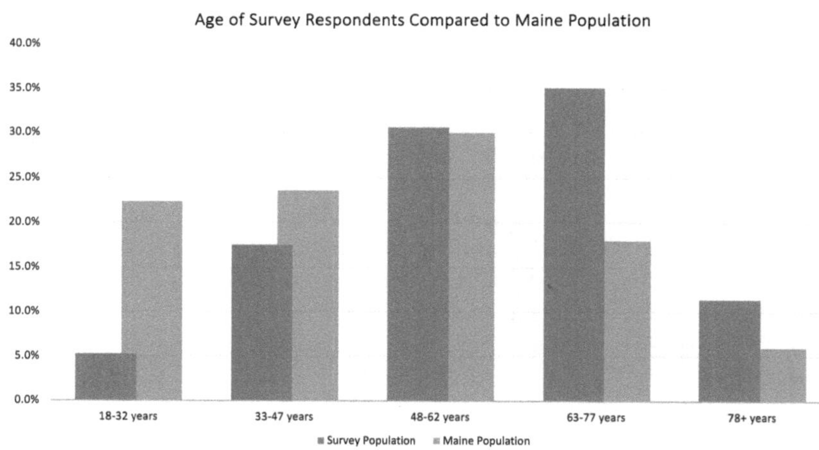

Fig. 3.2 Age of survey respondents compared to Maine population

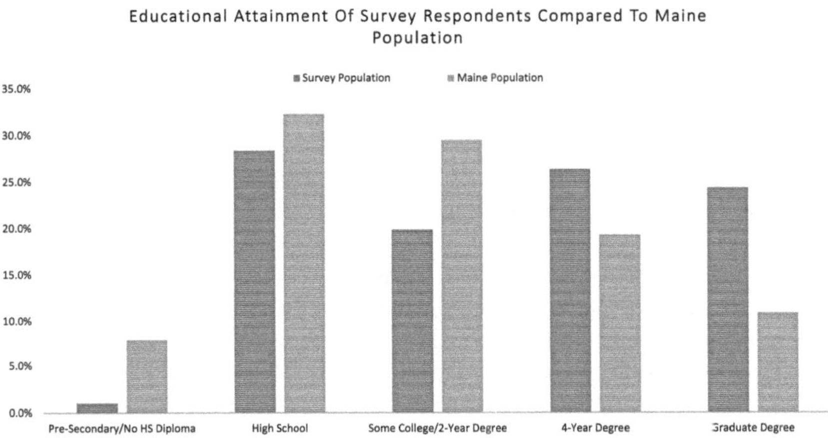

Fig. 3.3 Educational attainment of survey respondents compared to Maine population

We undertook an exploratory quantitative analysis to ascertain empirical relationships between reuse practices, socio-demographic characteristics of respondents, and social capital measures. Our survey design allowed us to make a distinction between the types of reuse activity respondents were involved in, and we created three dichotomous variables: "sell reuse," "buy reuse," and "free reuse." To generate the "sell reuse" variable, individuals were included if they reported selling goods over the past 12 months through a range of options (e.g. yard/tag/garage sale, online or print catalog, electronic platforms like Craig's List or Facebook, or through a national resale organization). To generate the "buy reuse" variable, individuals were included if they purchased used items over the past 12 months through a range of options (e.g. stores: consignment, thrift, charity, flea market, salvage yard, garage/tag/yard sale, online listing, etc.). Finally, we created a "free reuse" variable that included individuals who, over the past 12 months, had participated in nonmonetized forms of reuse exchange (e.g., donations to or received from charity stores, given to friends or family, found at transfer stations or the curb, community swap, etc.)

To explore the effect of social capital in the survey, we relied on four dimensions of social capital: personal relationships, social network support, civic engagement, and trust and cooperative norms. Initially we grouped social capital metrics into these categories of trust, social networks, personal relationships, and civic engagement. We then ran separate factor analyses on each grouping of variables. After using a varimax rotation, the loading scores combined with the cronbach alpha, we decided to include some of the raw variables instead of using the composites. From our analysis on trust for example, two out of the eight variables were found to fall into a common dimension. To this end, instead of losing the individual variation from those variables in a composite measure they were included separately in the model to preserve this information. It was also discovered that the personal relationships category proxied by the variable "number of close friends" appeared to be truncated in the final sample used for the model, to this end it was dropped.[2] Table 3.1 provides a breakdown of the dimensions of social capital, the variables used to measure them, and how they are connected to social capital in the literature.

Given the formulation of the data—three dichotomous dependent variables—three probit models were constructed and estimated using maximum likelihood. The models also included control variables for gender, age and income (panel A of Table 3.2) and the measures of social capital (panel B of Table 3.2). To interpret the probit coefficient estimates, average marginal effects were also estimated for all of the models (Table 3.3).

Qualitative Methods: Ethnography of Networks and Things in Motion

We undertook qualitative research to explore relationships in context, following suggestions in the literature that indicate the need for the context and nuance that qualitative data can provide (Grootaert and van Bastelaer 2002). We focused on the movement of goods through networks structured by relationships. We asked participants where they went to obtain used goods and how they got rid of things they no longer wanted or needed. We asked thrift, vintage, and secondhand retailers where they got objects and who bought from them. These discussions led us to create a map of the movement of goods, which also serves as a map of the social relationships that tie people and organizations to each other (see Fig. 3.4).

Our ethnographic fieldwork was based in secondhand economies in two rural communities: Mill Town and neighboring Jordan, but the work extends out from there, to nearby towns, larger cities, across the state of Maine, and into global networks of exchange. From May to December 2019 the lead author engaged in a wide range of reuse practices, from

Table 3.1 Dimensions of social capital and approach to measurement

Social capital dimension	*Definition*	*Variables*	*Literature*
Personal relationships	"web of connections between a given individual and the people that he or she knows" (Scrivens and Smith 2013, 21)	About how many close friends do you have these days? These are people you feel at ease with, can talk to about private matters, or call on for help.	(Scrivens and Smith 2013; Putnam 2000b)
Social network support	"the level of resources or support that a person can draw from their social contacts" (Scrivens and Smith 2013, 25)	If someone in my community is in need (death in family, lost job) people will pitch in to help out. If my community faces a common problem, people will cooperate to solve it.	(Grootaert and van Bastelaer 2002; Scrivens and Smith 2013; Putnam 2000b)
Civic engagement	"actions and behaviours that can be seen as contributing positively to the collective life of a community or society" (Scrivens and Smith 2013, 26)	Approximately how many hours a month, on average, do you volunteer?	(Hofferth and Iceland 1998; Scrivens and Smith 2013; Putnam 2000b)
Trust & cooperative norms	the "way people behave towards each other and as members of society" (Scrivens and Smith 2013, 34)	How much do you trust that those you are exchanging with are giving you a fair deal?—INDIVIDUAL SELLER I DO KNOW How much do you trust that those you are exchanging with are giving you a fair deal?—INDIVIDUAL SELLER I DO NOT KNOW Generally speaking, most people can be trusted. People selling used goods in my community will try to take advantage of others.	(Grootaert and van Bastelaer 2002; Putnam 2000a, b)

Table 3.2 Controls and social capital measures

Variable	*Obs*	*Mean Std.*	*Dev.*
Panel A: Control variables			
sell_reuse	531	0.4105461	0.4923968
buy_reuse	*531*	0.8380414	0.3687602
free_reuse	531	0.6685499	0.4711783
gender	508	1.596457	0.4910915
age	527	3.235294	1.049011
education	521	3.439539	1.162181
income	531	9.448211	4.021951
children_n~r	531	0.4934087	1.004449
Panel B: Measures of social capital			
vol_hrs	514	2.118677	1.205705
Trust seller I know	436	2.224771	0.7396372
Trust seller don't know	454	3.453744	0.7406945
Trust people generally	*511*	2.951076	0.5666844
Trust sellers local community	506	3.025692	0.520217
Trust national non profits	493	2.521298	0.6641615
Trust exchanges online	398	2.738693	0.8588319
Trust in-person exchanges	408	2.279412	0.8501294
Support local community people in need	443	3.090293	0.6800124
Support local community working together	494	1.98583	0.5853187

volunteering in local and national thrift stores, a local service club, and a home goods bank to visiting yard sales, antique shops, book sales, and charity events. Our team also conducted semistructured interviews (Bernard 2011) with volunteers, managers, and board members of reuse organizations, as well as with individuals buying, selling, and browsing for used goods—both in person and online (n = 54). These interviews were recorded, and lasted between 40 minutes and 2 hours. Interviews were inductive (Svendsen 2006), allowing us to follow lines of inquiry as they emerged. We transcribed all interviews verbatim to capture the nuances of what was said (Kowal and O'Connell 2014). Field notes and interview transcripts were coded in NVivo 12 for Mac (QSR International Pty Ltd. 2018) using a structured coding system based on the four dimensions of social capital: personal relationships, social network support, civic engagement, and trust and cooperative norms.

Table 3.3 Marginal effects

	Margins	*Margins*
	Free-reuse	*Buy-reuse*
gender	0.00832	0.0729**
	−0.0511	−0.0282
age	−0.0927**	−0.0193
	−0.031	−0.0177
education	0.0146	−0.0232
	−0.0228	−0.0133
income	−0.0128	−0.00405
	−0.00735	−0.00433
vol_hrs	0.015	0.00691
	−0.0216	−0.0119
children_number	0.0317	0.0506*
	−0.0302	−0.0211
Trust seller I know	−0.0173	0.0517*
	−0.0416	−0.0238
Trust seller don't know	0.101**	−0.0171
	−0.0358	−0.0196
Trust people generally	0.142*	−0.0586
	−0.0604	−0.0355
Trust sellers local community	0.0744	0.0243
	−0.0614	−0.0361
Trust national non profits	−0.0322	0.0177
	−0.0336	−0.0186
Trust exchanges online	−0.0589	−0.0236
	−0.0396	−0.0222
Trust in-person exchanges	0.101**	0.0611*
	−0.0417	−0.024
Support local community people in need	0.0981	0.046
	−0.0509	−0.0283
Support local community working together	0.00357	−0.0143
	−0.0487	−0.0267
N	346	346

*$p < 0.05$, **$p < 0.01$, ***$p < 0.001$

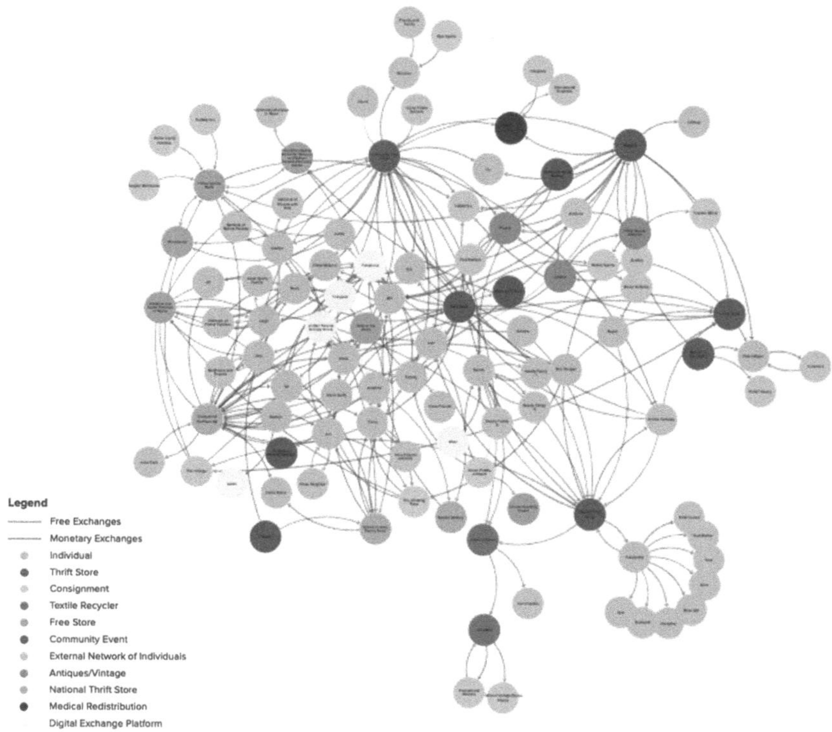

Fig. 3.4 Network map of reuse relationships

Results

Quantitative Results: Measuring Social Capital

Survey responses revealed that the overwhelming majority of participants (86 percent) had bought used goods or participated in nonmonetary reuse exchanges (84 percent), while nearly half (45 percent) had sold used goods over the past 12 months.

Table 3.4 displays the results of the probit estimations for each of the specified models. Our analysis of the relationships between social capital indicators and reuse indicates some commonalities across forms of participation. For example, people who participate as buyers and in nonmonetized exchanges report high levels of trust for their in-person exchange

Table 3.4 Probit estimations for social capital models

	Model 1	*Model 2*	*Model 3*
	Free-reuse	*Buy-reuse*	*Sell-reuse*
gender	0.027	0.535*	−0.078
	−0.16	−0.21	−0.15
age	−0.298**	−0.141	−0.122
	−0.1	−0.13	−0.09
education	0.047	−0.17	−0.079
	−0.07	−0.1	−0.07
income	−0.041	−0.03	0.02
	−0.02	−0.03	−0.02
vol_hrs	0.048	0.051	0.056
	−0.07	−0.09	−0.06
children_number	0.102	0.371*	0.106
	−0.1	−0.17	−0.08
Trust seller I know	−0.056	0.379*	0.07
	−0.13	−0.17	−0.12
Trust seller don't know	0.323**	−0.126	0.088
	−0.12	−0.14	−0.11
Trust people generally	0.456*	−0.43	−0.133
	−0.2	−0.26	−0.16
Trust sellers local community	0.239	0.178	−0.069
	−0.2	−0.27	−0.18
Trust national non profits	−0.103	0.13	0.008
	−0.11	−0.14	−0.1
Trust exchanges online	−0.189	−0.173	0.115
	−0.13	−0.16	−0.11
Trust in-person exchanges	0.326**	0.448**	−0.032
	−0.13	−0.17	−0.12
Support local community people in need	0.315	0.338	0.048
	−0.16	−0.21	−0.15
Support local community working together	0.011	−0.105	0.205
	−0.16	−0.2	−0.14
constant	−0.003	−0.063	−0.517
	−0.77	−0.95	−0.69
chi2	45.758	47.066	21.928
r2_p	0.114	0.187	0.046
N	346	346	346

$^{*}p < 0.05$, $^{**}p < 0.01$, $^{***}p < 0.001$

partners (significant, positive effect). However, there are also some differences as we look at social capital indicators and forms of participation in reuse. For example, survey respondents who reported participating in nonmonetized forms of reuse were more likely to trust in sellers they don't know.

Our examination of the marginal effects from our model suggests that, on average, for participants who reported buying secondhand items over the past 12 months—for each unit increase in their trust in sellers they don't know, there is a 5 percent increase in their likelihood of buying secondhand goods. Even more telling, we found that people who participated in nonmonetized exchanges during the year prior to the survey were 10 percent more likely to do so with each additional unit of trust they held for both in-person exchange partners and those that they did not know. We also found that those who reported participation in nonmonetized forms of reuse in the year prior to the survey were more likely to have been motivated to participate in order to support people in the community in need. For each one unit increase in their motivation to help others, there is a 9 percent increase in participating in nonmonetized reuse exchanges.

Overall, the survey results suggest that there are some positive connections between participation in reuse exchanges and indicators of social capital, most notably related to trust. These relationships appear to be stronger for people who report participating in nonmonetized forms of reuse. Notably, however, our quantitative analysis does not find strong relationships between reuse practices and other dimensions of social capital, including social network support, civic engagement, and personal relationships. Yet as our qualitative analysis will show, there are indeed strong relationships between all social capital indicators and participation in reuse economies—they simply resist quantitative measurement.

Qualitative Results: Impact(s) of Social Capital

Broadly speaking, our ethnographic work makes clear that there is much more to be understood about reuse and social capital than simply its presence or absence as indicated by metrics like trust. Indeed, in what follows we articulate the diverse impacts of social capital as we came to understand them through ethnographic engagement and qualitative data analysis. To illustrate the role of social capital in Maine's reuse markets we incorporate direct quotes from our interviews as well as ethnographic examples from our observations. While these illustrations provide solitary examples, they are representative of broadly shared patterns observed through months of participant observation and more than 50 interviews.

Personal Relationships

Examples of personal relationships forged through reuse exchanges were pervasive throughout our research. These examples included both deep and long-standing exchange relationships as well as fleeting, one-time transactions. For example, Peter is a bookseller who also engages in a local library-related charity. Although he no longer runs a brick-and-mortar bookshop, his retirement looks a lot like an active business buying and selling used books. He articulates his motivation for staying involved in the business, stating that "I still have customers I've had for 30 years and more. They keep me alive" (Interview, 8/28/19). He can trace the lineage of his books through their owners, and some books even come back to him when their owners die. Indeed, for Peter, the books seem to be a means of connecting people to each other.

These relationships were not secondary to reuse practices, but instead, for many, they were the primary reason for participation in this economic practice.

These ties bind people through close, enduring connections but there are also "weak ties" (Granovetter 1973) present in reuse economies. These weak ties between acquaintances can lead to connections that link people to broader networks. James, an antique shop owner, describes the presence of weak ties that seem to emerge through the connection the used objects provide:

> Every single thing in here that somebody buys is something that I found. You know what I mean? [...] I wonder about who's played with this toy or who treasured it, because especially if they're things, for example that 18th century bowl. That's been somewhere. It wasn't in the ground. Someone has cared for that bowl since the 1770s, and cared for it enough to pass it on and pass it on and pass it on. And so all of these things survived compared to the things that didn't. So there was a reason somebody kept them and treasured them. So I wasn't the first one. (Interview, 9/16/19)

Personal relationships forged through reuse, however, are not uniformly positive. Delia is a social worker who often helps her clients access used goods. She describes an experience she had with a thrift shop manager in Mill Town:

> I haven't been treated as nicely as I would've liked...I actually got yelled at one day myself by the guy that's running the place [...] I was not happy with that. I let him know that. And they're equally disrespectful to clients. (7/2/19)

Many of her clients struggle with mental health, are—or were—without access to permanent shelter, and are extremely poor. Delia's experiences demonstrate that some individuals are kept from the relational world of reuse. Instead of the family-like atmosphere described by some of our participants, Delia explains the web of bureaucratic processes that her clients must navigate to receive help from larger thrift shops. This bureaucracy is a means of removing personal relationships from reuse:

> So you can use the thrift store but you have to meet them at the Welfare Office. [That] is what they call the social services office. You have to meet with a lady there, tell them about your income, give them personal information, and you have to have somebody with you to go to the thrift store [...] they'll give you like ten items. It's not a lot. As long as you're income-eligible they will give you these ten items and that's all you get for three months [...] You can get some basic household items, but not a lot [...] they weren't even offering that unless it was an emergency. So you had to have someone vouch for you that it was an emergency. Like homeless is an emergency, or you know, you just got displaced from a fire and you need clothing. (Interview, 7/2/19)

If personal relationships enliven reuse for some participants, it is clear that these relationships are not equally available to all. Class, race, and housing status delimit who can build personal relationships and who must conform to institutional processes to receive a predetermined level of support for approved emergencies.

Social Network Support

A key indicator of social capital, utilized by both public good and critical scholars are social networks. We found overwhelming evidence of social network support and instances when social networks were exclusive within reuse practices. Many reuse organizations we spoke and worked with were small-scale, run totally by volunteers. For these organizations, drawing on social network support was critical to their operations. Jon, the founder of a local home goods bank, relies on people to give what they can to enable him to redistribute used goods to people in need. This means that he calls on his network of connections to find volunteers, but he also leverages relationships to gain access to needed equipment, like moving trucks for when he finds a large haul of goods. Jon describes how the moving company volunteers to help, at no charge, explaining that "if it's anything larger—like from a hotel or house being emptied out—we have a moving company...god love them. I love them. I have no idea how much they've actually spent on us. They'll never tell us" (Interview, 6/17/19).

We can also see evidence of social network support when people in the community face catastrophic life disruptions. When Ivan's apartment building burned down, he turned to a local thrift shop for clothing and necessities:

> We spend the night in the hotel and then the next morning it's like okay... my brother's in his pajamas. So I showed up there just for a pair of pants and they gave me, you know, it was like lots of clothes, toothpaste, tooth brush, they were scurrying around putting together this little care package for me. And it was just like...it really impressed me [...] as soon as they realized what my situation was they all set to work, you know? And they were not going to let me leave without doing as much as they possibly could for me. And that's like—that's very un-Wal-Mart like. (Interview, 8/19/19)

These examples of social network support reflect a vision of social capital that is useful, productive, and *good*. In these instances, the collective efforts of friends, neighbors, and community institutions allow people to get by and to thrive, even during difficult times. Yet these positive outcomes must be countered by an understanding of how social network support can also exclude.

The very networks that allow some individuals to feel warm and welcomed by thrift shops, like Ivan, can make others feel unwelcome. Delia, the social worker whose clients are often unhoused and/or suffering from mental health issues, finds the same shop Ivan visited extremely off-putting. She notes that "they would prefer to only give to their church members. And I think they kind of look down on others that come there. They give you very little" (Interview, 7/2/19). The strong bonds that link some to this shop keep others out.

For many of the individuals we spoke with, connecting people "in need" to used goods was important. Yet determinations about how to define "need" were rooted in cultural assumptions about worthiness. For example, Carol, the manager of a local thrift shop, prides herself on knowing the community—who needs help and who doesn't—and will give her support to those who don't ask for it...those who are not looking for "handouts" (Field Notes, 8/20/19). This practice emerged in other thrift shops, as well. Identifying those who were *truly* in need was thought to be a skill that volunteers could acquire—learning "who to really soak it to [charge extra] and who to give it to [for free]" (Interview, Jolene, 10/14/19). Yet in making these judgments, volunteers imparted their own notions of deservedness that likely excluded many people who could have benefitted from access to opportunities or objects.

Civic Engagement

In the context of reuse, contributing positively to the collective life of a community often took the form of volunteering in reuse organizations or donating used goods. From this perspective alone, the link between community engagement and reuse is quite apparent in our ethnographic work. Jill, a frequent volunteer at a thrift shop in Jordan, describes the value of her work for the community, saying "I feel really gratified that I've used my ability to make money for this organization that does so many good things" (Interview, 8/14/19). Indeed, apart from giving individuals access to goods at low- or no cost, many reuse organizations use their proceeds to fund programs and services for people in need. Two local thrift shops in our study communities generate more than $40,000 each per year, funneling this money into community suppers, local library and school programs, outreach and support for people in need, and critical services like a local ambulance, food pantry, and more. Civic engagement, then, is a motivator for people to participate in reuse and an outcome of that participation.

Civic engagement can also be problematic. While Jill sees the value in her work at the thrift shop, she goes on to say that "There are times when I resent it, because I do work so hard and other people don't" (Interview, 8/14/19). This volunteer work asks a lot of individuals, creating emotional and physical burdens (Berry 2022) for the small group of people who are willing to give their time to help their communities. Localized reuse is heavily reliant on volunteer labor (Berry 2022), yet volunteers are aging and are not being replaced by younger residents. Lindsey, a volunteer with a local service club that runs an annual community yard sale, explains almost despondently that community members are "willing to give you money, but they aren't willing to give you time" (Interview, 7/20/19). He says that the sale, which has been a community tradition for decades, is on the decline. As volunteers age, and become increasingly unable to do the hard work associated with reuse the reliance on volunteer labor may be problematic (Berry 2022). This reality echoes critiques of social capital theory based on its tendency to push self-sufficiency and provision of services onto volunteer labor (Naughton 2014).

Trust and Cooperative Norms

A final element of social capital, trust, emerges in reuse economies in a number of forms. The heterogeneity of used goods means that establishing their value is exceedingly challenging. Many thrift shop managers and volunteers struggle to determine the price of objects. Volunteer organizers of a community sale told us about a patron who bought a used violin from the sale for $40. He later returned the violin to the organizer of the sale, saying only "you should get this appraised." The organization took his advice, and later re-sold the violin for $1500 to another buyer. Obtaining a fair price for objects is reliant on trust between those who know the true worth of goods and those who don't.

On some occasions, reuse organizations forgo the opportunity to get a fair price for goods intentionally, as Ivan describes:

> I had this big pile of stuff at the thrift store and it's like "okay, there's my pile. I can't find anything more to buy. It's time to check out." I discovered I only have $4 bucks in my wallet and I've got like $20 worth of stuff there. And the lady predictably says "oh...just..." and I was like "no, I can't do... well here's my $4 bucks, but here's what I'll do...." And I noticed there was a community supper there. So I showed up for the community supper and like made a $20 donation. (Interview, 8/19/19)

Here the thrift shop volunteer trusts that Ivan truly only has $4 in his wallet. This demonstrates trust in the general community—whether people in the community are worthy of help and support. Many thrift and reuse shop owners described allowing customers to take objects without payment. Anna owns a small secondhand shop, and told me that "if somebody comes in and they need something and they didn't have any money...I'm like well take it and bring it when you get some money" (Interview, 6/20/19). This trust in the community allows people to benefit from access to used goods.

Yet if there are those in the community who benefit from implicit trust, there are others who do not. Sam, who works with Indigenous survivors of domestic violence, describes with a mix of humor and frustration how her clients are surveilled by volunteers when they attempt to access free used goods from a local reuse non-profit:

> the volunteers [...] will follow you around and be like "oh, you can only take two rolls of toilet paper" or whatever. And people that we work with—they're intimidated by being followed. They already feel like "ugh, I'm going and getting stuff for free" and we want them to be as comfortable as they can, and we don't want them to feel like they're begging or getting hand-me-downs or whatever. (Sam, 7/10/19)

Even as they attempt to redistribute goods at no cost, volunteers express a lack of trust in the people who attempt to access items. The people Sam works with are often Indigenous and poor, indicating that trust in "the community" indicates trust in a certain definition of community. Who is trusted to pay later? Who seems deserving of access to goods despite an inability to pay? These questions indicate that trust in reuse economies brings benefits to some, but certainly excludes others across boundaries that can be racialized and classed.

Discussion

Our work certainly supports the suggestions that reuse is rooted in sociality (Barnett et al. 2005; Carrier and Luetchford 2012), with potential benefits for well-being (Grobsmith 1981) and for creating localized networks of reciprocal support (Isenhour 2003). Our exploratory quantitative analysis clearly establishes relationships between reuse practices and components of social capital—most notably the dimension of trust. Yet our qualitative analysis urges us to consider the diverse impacts of social capital associated with reuse. For if we blind ourselves to the repercussions of dense, tight networks of relationships, we see a falsely positive social capital. Stories of exclusion, of worthiness (narrowly defined), or a lack of worthiness for help (broadly defined), of trust broken, or of strains on civic leaders were also common in the ethnographic portions of our project. Here we see a social capital that excludes *and* includes. It elevates some at the expense of others. This project exposes some of the gaps in common understandings of social capital—highlighting how quantitative measures, when used alone and in abstraction, fail to capture social capital as it is experienced by individuals.

Scholars have lamented the difficulty of measuring social capital (Portes 1998; Warde and Tampubolon 2002) and critiqued it as a "chaotic concept" that has been "colonized" by economics (Fine 1999). J.K. Gibson-Graham argue quite compellingly that social capital is based on a "capitalocentric assumption [...] that the social relations addressed by these concepts are 'investments' that can eventually be monetized, exchanged, and used to generate profitable returns" (Gibson-Graham 2006, 59). Yet in exploring the results here, we ask whether social capital might help us rethink the concept of capital altogether. A social capital that is evaluated in both quantitative and qualitative terms can be read as deeply relational and entangled in power and politics. It need not be another tool for economic colonization, but instead a means for exploring the complex web of social relationships that power local economies.

Conclusion

Rejecting the too-easy narrative that secondhand exchanges are the answer to (re)building community relationships, this paper has explored how the movement of used goods can be inclusive and exclusive with respect to diverse members within the communities where we worked. This understanding is critical as we seek solutions to issues like global resource use, climate change, and waste. Secondhand economies are an important part of a more circular economic system that reduces waste by extending the lifespans of consumer goods (EMF 2013). Some proponents of more circular economic systems argue that such systems will (re)build social capital, presumably with positive impacts anticipated for all. Our work suggests that reuse is indeed related to social capital, but that our understanding of social capital must become significantly more complex if we are to account for social capital as it is experienced rather than as it is often theorized. Our analysis demonstrates the need for mixed-methods and qualitative research to explore social capital, rather than using quantitative methods alone.

Just as the public goods scholars have tended to romanticize the positive impacts of social capital, some proponents of reuse economies, collaborative consumption, and sharing economies have sought to highlight the positive social impacts without as much attention to the potential negative impacts (e.g., Botsman and Rogers 2010). Here we urge caution. While reuse economies have the potential to generate all sorts of value (Isenhour and Berry 2020), fairness, equity, and inclusion are not guaranteed—they must be designed into programs and policies meant to support resource conservation and community well-being (Berry et al. 2021). Understanding reuse as a relational process that is rife with care and connection as well as exclusion and power might mean that there is a need for diverse, localized exchanges that are sensitive to existing personal networks. This approach might allow us to see circular practices, like reuse, as complex social processes rather than as policy end-points.FundingThis work was supported by the National Science Foundation Cultural Anthropology Program (Grant 1756933).

Notes

1. People, places, and organizations described in this chapter have been assigned pseudonyms to protect their privacy.
2. Initially in model construction the personal relationships variables were included but given the little variation in the data it added little to the analysis and as such was dropped for the final runs.

References

Adger, W. Neil. 2003. Social Capital, Collective Action, and Adaptation to Climate Change. *Economic Geography* 79 (4): 387–404. https://doi.org/10.1111/j.1944-8287.2003.tb00220.x.

Aldrich, Daniel P., and Michelle A. Meyer. 2014. Social Capital and Community Resilience. *American Behavioral Scientist* 59 (2): 254–269. https://doi.org/10.1177/0002764214550299.

Apinunmahakul, Amornrat, and Rose Anne Devlin. 2008. Social Networks and Private Philanthropy. *Journal of Public Economics* 92 (1): 309–328. https://doi.org/10.1016/j.jpubeco.2007.07.005.

Aytac, Deniz Ozge, Tulin Vural Arslan, and Selen Durak. 2016. Adaptive Reuse as a Strategy Toward Urban Resilience. *European Journal of Sustainable Development* 5 (4): 523–523. https://doi.org/10.14207/ejsd.2016.v5n4p523.

Barnett, Clive, Nick Clarke, Paul Cloke, and Alice Malpass. 2005. The Political Ethics of Consumerism. *Consumer Policy Review* 15 (2): 45–51.

Benevolenza, Mia A., and Lea Anne DeRigne. 2019. The Impact of Climate Change and Natural Disasters on Vulnerable Populations: A Systematic Review of Literature. *Journal of Human Behavior in the Social Environment* 29 (2): 266–281. https://doi.org/10.1080/10911359.2018.1527739.

Bernard, H. Russell. 2011. *Research Methods in Anthropology: Qualitative and Quantitative Approaches*. 5th ed. Lanham, MD: AltaMira Press.

Berry, Brieanne. 2022. Glut: Affective Labor and the Burden of Abundance in Secondhand Economies. *Anthropology of Work Review*. https://doi.org/10.1111/awr.12233.

Berry, Brieanne, Brianna Farber, Fernanda Cruz Rios, Michael A. Haedicke, Sujan Chakraborty, Sara Sophia Lowden, Melissa M. Bilec, and Cindy Isenhour. 2021. Just by Design: Exploring Justice as a Multidimensional Concept in US Circular Economy Discourse. *Local Environment* 1–17. https://doi.org/10.1080/13549839.2021.1994535.

Borgonovi, Francesca, and Elodie Andrieu. 2020. Bowling Together by Bowling Alone: Social Capital and COVID-19. *Social Science & Medicine* 265 (November): 113501. https://doi.org/10.1016/j.socscimed.2020.113501.

Botsman, Rachel, and Roo Rogers. 2010. *What's Mine Is Yours*. New York: Harper Business.

Bourdieu, Pierre. 2004 [1986]. The Forms of Capital. In *The Routledge Falmer Reader in Sociology of Education*, ed. S. J. Ball, 15–29. New York: Routledge Falmer.

Carrier, James G., and Peter Luetchford, eds. 2012. *Ethical Consumption: Social Value and Economic Practice*. 1st paperback ed. New York: Berghahn Books.

Coleman, James S. 1988. Social Capital in the Creation of Human Capital. *American Journal of Sociology* 94:S95–S120.

Demetry, Daphne, Jessica Thurk, and Gary Alan Fine. 2015. Strategic Poverty: How Social and Cultural Capital Shapes Low-Income Life. *Journal of Consumer Culture* 15 (1): 86–109. https://doi.org/10.1177/1469540513493205.

Duncan, Cynthia M., and Angela Blackwell. 2014. *Worlds Apart: Poverty and Politics in Rural America*. 2nd ed. New Haven, London: Yale University Press.

Durrenberger, E. Paul. 2002. Commentary: Why the Idea of Social Capital Is a Bad Idea. *Anthropology News* 43 (9): 5–5. https://doi.org/10.1111/an.2002.43.9.5.1.

EMF. 2013. *Towards the Circular Economy Vol. 2: Opportunities for the Consumer Goods Sector*. Ellen Macarthur Foundation. https://ellenmacarthurfoundation.org/towards-the-circular-economy-vol-2-opportunities-for-the-consumer-goods.

EMF. 2022. FAQ: What Is the Circular Economy? https://ellenmacarthurfoundation.org/media-centre/overview. Accessed 19 July.

Fine, Ben. 1999. The Developmental State is Dead—Long Live Social Capital? *Development and Change* 30 (1): 1.

Flora, Cornelia Butler, Jan L. Flora, and Stephen P. Gasteyer. 2015. *Rural Communities: Legacy and Change*. 5th ed. Boulder, CO: Westview Press.

Gallent, Nick. 2014. The Social Value of Second Homes in Rural Communities. *Housing, Theory and Society* 31 (2): 174–191. https://doi.org/10.1080/14036096.2013.830986.

Gibson-Graham, J. K. 2006. *Postcapitalist Politics.* Minneapolis, MI: University of Minnesota Press. http://ebookcentral.proquest.com/lib/umaine/detail.action?docID=310738.

Granovetter, Mark. 1973. The Strength of Weak Ties. American Journal of Sociology 78 (6): 1360–80.

Grobsmith, Elizabeth. 1981. *Lakota of Rosebud: A Contemporary Ethnography.* Fort Worth, TX: Harcourt Brace Johanovich Publishers.

Grootaert, C., and T. van Bastelaer. 2002. *Understanding and Measuring Social Capital: A Multidisciplinary Tool for Practitioners.* Washington, DC: World Bank.

Hamilton, Matthew L., and Mark Lubell. 2019. Climate Change Adaptation, Social Capital, and the Performance of Polycentric Governance Institutions. *Climatic Change* 152 (3): 307–326. https://doi.org/10.1007/s10584-019-02380-2.

Hanifan, Louis J. 1916. The Rural School Community Center. *The Annals of the American Academy of Political and Social Science* 67 (1): 130–138. https://doi.org/10.1177/000271621606700118.

Hofferth, S. L., and J. Iceland. 1998. Social Capital in Rural and Urban Communities. *Rural Sociology* 63 (4): 574–598. https://doi.org/10.1111/j.1549-0831.1998.tb00693.x.

Isenhour, Cindy. 2003. *Following the Buffalo: Lakota Value Construction and Consumption in Shifting Economic Environments.* Master's thesis. Colorado State University.

Isenhour, Cindy, and Brieanne Berry. 2020. "Still Good Life": On the Value of Reuse and Distributive Labor in "Depleted" Rural Maine. *Economic Anthropology* 7 (2): 293–308. https://doi.org/10.1002/sea2.12176.

Jones, Eric C., Arthur D. Murphy, A. J. Faas, Graham A. Tobin, Christopher McCarty, and Linda M. Whiteford. 2015. Postdisaster Reciprocity and the Development of Inequality in Personal Networks. *Economic Anthropology* 2 (2): 385–404. https://doi.org/10.1002/sea2.12038.

Jordan, Joanne Catherine. 2015. Swimming Alone? The Role of Social Capital in Enhancing Local Resilience to Climate Stress: A Case Study from Bangladesh. *Climate and Development* 7 (2): 110–123. https://doi.org/10.1080/17565529.2014.934771.

Kowal, Sabine, and Daniel C. O'Connell. 2014. Transcription as a Crucial Step of Data Analysis. In *The SAGE Handbook of Qualitative Data Analysis*, ed. Uwe Flick, 64–78. Thousand Oaks, CA: SAGE.

Leahy, Jessica E., and Dorothy H. Anderson. 2010. "Cooperation Gets It Done": Social Capital in Natural Resources Management Along the Kaskaskia River. *Society and Natural Resources* 23 (3): 224–239. https://doi.org/10.1080/08941920802378897.

Lin, Nan. 1999. Social Networks and Status Attainment. *Annual Review of Sociology* 25 (1): 467–487. https://doi.org/10.1146/annurev.soc.25.1.467.

Malecki, Edward J. 2012. Regional Social Capital: Why It Matters. *Regional Studies* 46 (8): 1023–1039. https://doi.org/10.1080/00343404.2011.607806.

Narotzky, Susana. 2007. The Project in the Model: Reciprocity, Social Capital, and the Politics of Ethnographic Realism. *Current Anthropology* 48 (3): 403–424. https://doi.org/10.1086/512999.

Naughton, Linda. 2014. Geographical Narratives of Social Capital: Telling Different Stories About the Socio-Economy with Context, Space, Place, Power and Agency. *Progress in Human Geography* 38 (1): 3–21. https://doi.org/10.1177/0309132513488731.

Pitas, Nicholas, and Colin Ehmer. 2020. Social Capital in the Response to COVID-19. *American Journal of Health Promotion* 34 (8): 942–944. https://doi.org/10.1177/0890117120924531.

Portes, Alejandro. 1998. Social Capital: Its Origins and Applications in Modern Sociology. *Annual Review of Sociology* 24 (1): 1–24. https://doi.org/10.1146/annurev.soc.24.1.1.

Prewitt, Kenneth, Christopher D. Mackie, and Hermann Habermann. 2014. *Civic Engagement and Social Cohesion: Measuring Dimensions of Social Capital to Inform Policy.* National Academies Press.

Putnam, Robert D. 1995a. Bowling Alone: America's Declining Social Capital. *Journal of Democracy* 6 (1): 65–78. https://doi.org/10.1353/jod 1995.0002.

Putnam, Robert D. 1995b. Tuning In, Tuning Out: The Strange Disappearance of Social Capital in America. *PS: Political Science and Politics* 28 (4): 664–683. https://doi.org/10.2307/420517.

Putnam, Robert D. 2000a. *Social Capital Benchmark Survey.* Roper Center for Public Opinion Research. https://ropercenter.cornell.edu/featured-collections/2000-social-capital-community-benchmark-survey.

Putnam, Robert D. 2000b. *Bowling Alone.* Simon and Schuster.

QSR International Pty Ltd. 2018. *NVivo* (version 12). NVivo Qualitative Data Analysis Software.

Rupasingha, Anil, Stephan J. Goetz, and David Freshwater. 2006. The Production of Social Capital in US Counties. *Journal of Socio-Economics* 35:83–101.

Sawhill, Isabel V. 2020. *Social Capital: Why We Need It and How We Can Create More of It.* Brookings Institution.

Schneider, Jo Anne. 2006. Anthropological Relevance and Social Capital. *Anthropology News* 47 (3): 4–5. https://doi.org/10.1525/an.2006.47.3.4.

Schneider, Jo Anne. 2008. Social Capital, Civic Engagement and Trust. *Anthropologica* 50 (2): 425–428.

Scrivens, Katherine, and Conal Smith. 2013. *Four Interpretations of Social Capital: An Agenda for Measurement.* OECD Statistics Working Papers. OECD Publishing. https://doi.org/10.1787/5jzbcx010wmt-en.

Svendsen, Gunnar L. H. 2006. Studying Social Capital In Situ: A Qualitative Approach. *Theory and Society* 35 (1): 39–70. https://doi.org/10.1007/s11186-006-6780-3.

Wall, Ellen, Gabriele Ferrazzi, and Frans Schryer. 1998. Getting the Goods on Social Capital. *Rural Sociology* 63 (2): 300–322.

Warde, Alan, and Gindo Tampubolon. 2002. Social Capital, Networks and Leisure Consumption. *The Sociological Review* 50 (2): 155–180.

Woolcock, Michael. 2001. The Place of Social Capital in Understanding Social and Economic Outcomes. *Canadian Journal of Policy Research* 2 (1): 65–88.

CHAPTER 4

Greening Farmers Markets: How Communities Come Together to Keep Things Out of the Landfill

Susan Andreatta

Identifying opportunities for communities to come together to address climate change and limit waste ending up in landfills is critical given the time horizon remaining to undertake significant activities to reduce global temperatures, CO_2, other greenhouse gas emissions, and conserve access to landfill space. The IPCC and other climate scientists argue there is less than a decade to lower the earth's temperature to protect it from uncertain climatic conditions (Bonnisseau 2022; UN 2021 and 2022). In North Carolina, where this chapter is placed, there are only 25 more years of landfill space remaining, barring no catastrophe (such as leaching and fires). In the face of these pressing circumstances, questions arise as to how the public can take action and responsibility to help the environment recover. What can individuals and communities do? Do their individual acts even make a difference? Can a community-based approach engender larger change and serve as an exemplar for other groups to follow?

S. Andreatta (✉)
Greensboro, NC, USA
e-mail: S_ANDREA@uncg.edu

P. Kuppinger (ed.), *Second-Hand Cultures and Economies of Reuse, Repair, Sharing, and Care*,
https://doi.org/10.1007/978-3-031-99875-1_4

This chapter examines two realms of everyday life where dedicated individuals and groups work to make a difference. I analyze how people seek to green farmers markets and how communities can come together in this context and make local environments more sustainable. I describe how the public's participation at a local farmers market helps to make a difference in greening the market experience, that is avoiding trash and shortening distances traveled. I also describe how a group of undergraduate students from a state university chose to modify their behavior for a semester and monitor their activities illustrating how their changed behaviors worked to lessen their environmental footprint and contributed to the community taking care of the earth. The ease with which these students and the farmers market participants modified their behaviors provides opportunities for others to do the same. I draw on ideas and debates from the degrowth movement to highlight pathways where the public, and here concretely farmers, vendors, community members, or students can reduce, reuse, repurpose, or recycle things. Working together, farmers, vendors, market managers, market volunteers, and community members foster community cohesion and engagement while also greening markets.

Farmers markets are frequently located in urban areas, where the built environment does not allow for much space to grow fresh produce. Farmers markets are places where choices can be made to minimize packaging, limit how far food travels, and ensure equitable access to local, sustainable products; they foster green consumption. I discussed these ideas and realities with students while introducing them to notions and discussion of degrowth. Students were then asked to change one behavior they thought would make a difference in the community's local environment during the course of a semester. Throughout this chapter, I insert examples from students' work and individual projects, highlighting those who reduced their use of plastic water bottles, drinking coffee, and limited their car use. These experiences show how a farmers' market and a class assignment transformed people's behavior and how they embraced different sustainable practices and aligned them with social and food justice. Using these examples, I showcase possibilities for individuals and communities to change behaviors to limit waste ending up in landfills and in other ways to contribute to the struggle against climate change. The examples of (further) greening aspects of a farmers' market and changing individual behaviors feed into larger economies of reuse, repair, sharing, and care and especially points to ideas of reducing waste as a contribution and act of care for a more sustainable city and world.

Degrowth

Originating in western Europe, degrowth is a concept, framework, or philosophy that was introduced over fifty years ago (D'Alisa et al. 2015). Since then, conferences, workshops, and courses have spread the concept and relevant debates across the world, including North and South America, Africa, and Australia. However, it has yet to garner wider appeal or more concrete action. It is understood among "degrowthers" or those aware of the concept that degrowth in the northern hemisphere will be and needs to be different from practices in many parts of the southern hemisphere.

Degrowth is defined from several perspectives. For many writers, thinkers, and practitioners, degrowth is a philosophy of life or lifestyle that calls for conscious efforts to reduce, reuse, recycle, and repurpose (Latouche 2009; Muraca 2013; Paulson 2017, 2020). Others understand degrowth to be part of a political and social movement based on ecological economics designed to lessen consumerism, production, and colonial resource extraction. Still others put forth an economic geographical perspective and challenge the dominant global economy that is based on ideas of continuous growth and development and is dependent on the use of finite nonrenewable resources, in particular, fossil fuels (Kallis 2015). Generally, degrowth critiques capitalism and growth for the sake of growth that prioritizes profits, production, and consumption as indicators of well-being over quality of life, societal well-being, and environmental health (D'Alisa et al. 2015; Kallis 2015; Kallis et al. 2018; Paulson 2017; Saitō 2020).

The current system of hyperconsumerism has accustomed consumers to convenience, specifically having their needs quickly and easily met. The marketplace encourages cultural values of convenient solutions (e.g., single-use plastics), which often compromise the environment in both the short and long term. From its ecological and economic stance, the degrowth movement calls for a shift away from capitalism-driven economic systems measured by gross domestic products (GDP; D'Alisa et al. 2015) and seeks alternatives. Included in this approach is the quest for a basic minimal income along with capping salaries to reduce income disparities, excessive wealth, and poverty.

Participants in the degrowth movement and its debates re-/envision ways of living that emphasize quality of life, well-being, care, and conviviality while serving as an economic strategy to respond to the limits-to-growth dilemma (Kallis 2015; Latouche 2004; Muraca 2013; Paulson 2017, 2020; Andreatta 2021). Degrowth calls for a restructuring of cultural values and of how "success" is measured. It calls for the inclusion of environmental damage and resource depletion into products and prices (Kallis 2015). It advocates downsizing and reducing overall material consumption and dependency on fossil fuels and provides a template for a society to consume and produce less, share and care more, and strive for higher levels of equity amongst people and nations. By reducing production and consumption practices, degrowth also fosters a reduction in greenhouse gas emissions connected to the transportation sector, seeking to engender a more sustainable global future (Muraca 2013; Kallis 2015; Kallis et al. 2018; Saitō 2020).

Curiously, for all that has been written on degrowth, the urgency and immediate need for it, what appears to be missing from the literature is how to effectively operationalize the concept in everyday life. There is no manual that suggests how to change policy from the top down either at the community or state level. There are no guidelines on how to measure the impact of degrowth on a collective scale using new indicators. The question emerges of how would one measure increased mindfulness or happiness in a community, state, or even globally? However, what could be measured is how much less waste landed in the landfill or if there was less food waste, and food and housing insecurity. We could measure if the earth's temperature began to decline along with CO_2, and other greenhouse gas emissions decreased at significant rates. At present our most urgent collective challenge is how to produce such an impact and how to do so as fast as possible. Is it a policy approach from the top down or an activists' project from the bottom up? Is it an economically driven approach, a philosophical one, or a cultural and behavioral one that would yield the most effective results? Many questions loom but making the impact systemic is paramount to transform and improve the ways in which humans collectively interact with and care for the Earth. Next, I explore the concrete role of a local farmers market and sustainable practices that address climate change and limit landfill waste collectively and bespeak care for people and the planet.

The Greensboro Farmers Curb Market

The Greensboro Farmers Curb Market (GFCM) is located in the city of Greensboro in Guilford County, North Carolina. The GFCM is one of three farmers markets within a ten-mile radius. The GFCM is the oldest and most established of these markets. Also in Guilford County, there is a state-operated market, the Piedmont Triad Farmers Market, and some independent farmers markets. In 2023, the Greensboro Metro Area population had about 300,000 inhabitants and was the third largest in the state. Almost 540,000 people resided in Guilford County (Greensboro North Carolina – Government 2023).

The GFCM was created to provide a venue for farmers, vendors, and visitors to come together to sell and buy fresh local produce. It is named after the original farmers market, which began on a curb on Commerce Place in downtown Greensboro in 1874. Since its founding, it has affectionately been known as the "Curb Market." Although the market has moved a few times in the city center, it was never far from its original location. For the past 61 years, the GFCM has been using the city's old armory. In 2024 the "Curb Market" will celebrate its 150th anniversary. Sellers at the farmers market must reside or have their business or farm within a 100-mile radius of the market. Pre-COVID the market had 100 vendors and attracted over 3000 visitors every Saturday. Since COVID both vendors and visitors declined in attendance; however, the numbers are increasing again. There is a noticeable increase in new and younger farmers joining the market vendors (Andreatta and Hoskins 2022a, b).

As with any community-based organization, the market's identity is dynamic, adaptive, and responsive. The former market manager, Lee Mortensen,[1] who had been in her position for nearly a decade, along with many farmers, vendors, and customers embraced the concept of greening the market in a seamless, nonthreatening way. Mortensen used an informal approach to green consumerism (Godin and Langlois 2021; Wagner 2017). Between 2014 and 2023, she worked to enhance the market and create a diverse, welcoming, and environmentally friendly venue. In 2023 the Curb Market was the recipient for the first City of Greensboro's Friends of Sustainability Award. To earn this recognition, Mortensen encouraged farmers, vendors, and visitors to limit the use of single-use plastics (bags, clamshells, and bottles), containers (i.e., for blueberries, strawberries, peaches), and cups and lids (coffee and tea). Together, market visitors took home fewer or none of these items, and subsequently reduced the number of such items going to the landfill from their market experience (for details see below). Creating a culture of mindful sourcing, gathering, and purchasing fresh local harvest or local artisan products (crafts and baked goods) illustrates how farmers markets are at the intersection for greening consumerism and community-engaged sustainability.

Example 1. Keeping it Local at the Farmers' Market
Small to medium-scale farmers rely on farmers markets around the country to sell their harvest and value-added products. Since these farmers are not big enough to compete with larger and industrial-scale farmers selling wholesale to supermarket chains, many such farmers keep their products local and sell them at farmers markets. For example, the Curb Market is a producer-only market, and all products must be grown or made within a 100 miles radius of the market, limiting driving and greenhouse gas emissions by keeping it local for all consumers, farmers, and vendors. In this setting, farmers and vendors are also able to maintain personal connections with their customers. Their products are not traveling far for their intended use, unless of course an item is gifted and sent a distance to a recipient.

To encourage local purchases around holiday times, the market advertises on their website. For example, for Valentine's Day they offer the following:

> The Market is the BEST place to get all-local gifts and goodies for your Valentine's, Galentine's, friends and loved ones! Shop for meal ingredients, sweets, flowers. cards, gifts, knitted items, potted plants, body-care products, wood workings, jewelry, candles and so much more! Staff pick: Make your sweetie a Market Valentine's *Box-of-Love* filled with variety of local products and support your favorite vendors! (Greensboro Farmers Curb Market 2024)

Another example for keeping it local through community engagement and fostering sustainability is the interest in popup farmers markets. In Guilford County, neighborhood markets and food sharing take place, and were noted during the pandemic. Those involved found creative ways to retool farmers markets to meet local needs for fresh produce and provide for those out of work while we were all shuttered in place (Heuer et al. 2020; Andreatta and Hoskins 2022a, b). These community engagement activities contributed to sustainability and benefitted human health, reduced greenhouse gas emissions, and lessened the need for outsourcing and the global transportation for some goods.

Example 2. Recycling, Reusing, and Reducing

Farmers and vendors have been encouraged by farmers market management and visitors to recycle and reuse containers. For example, farmers selling plant transplants are willing to recycle and reuse plant containers and ask their customers to bring them back when they are finished, to limit what goes into the landfill. One farmer advertises with "Bring Back your Pots! We recycle any pot or tray we sell." Such reuse also reduces costs for farmers and vendors allowing them to buy smaller quantities of containers.

When amounts and units per sale matter, keeping packaging materials as minimal as possible and biodegradable is the way these farmers are heading. Some attempt to reduce the use of single-use plastics (bags, clamshells, cartons, and bottles), others use repurposed green pressed cardboard containers for strawberries and other berries and fruits and encourage customers to return these for reuse or refill upon their next visit. Similarly, those selling eggs reuse egg cartons. Some vendors reduce the price for a reused and refilled carton by twenty-five cents. Likewise, farmers sell blueberries in sturdy plastic clamshells that can be washed and returned for reuse. For example, Terry, from Rivers Finest Blueberries promotes recycling clamshell containers and baskets they use to sell their blueberries, along with jam, marmalade preserve, and honey jars. Terry always says thank you for each item a customer returns, acknowledging their effort. I broached Terry about taking back honey jars (they have 30 hives that help with blueberry production). She said because they currently used plastic containers, she would not feel comfortable reusing them for honey. However, her husband had not yet ordered jars for this honey season, and she said would speak to him about buying glass jars since so many of her preserve jars are returned. She also commented that I was not the first person to offer to return honey jars, so she thinks they will do it. By the time summer rolled around, Terry shared that her husband agreed to invest in both glass jars and plastic ones. The cost to sell honey in glass is higher than plastic and impacts the price. Terry explained, "we want to make our honey affordable to all, not only those who can just afford the glass" (Personal Communication 2023). Their next step is advertising that they will reuse the glass jars to encourage customers to return them and reduce waste.

Recycling takes place in several forms. The Curb Market offers a quarterly event called "Curiosities at the Curb," which is an upcycle shopping event tailored to selling vintage items, giving them a second life in a new home. Vintage finds include retro treasures, collectibles, and antiques. A share of the sales proceeds goes to support the market. The recycling educators from the city's local recycling center set up a booth at least six times a year. These educators promote recycling, reducing consumption, and more selective biodegradable purchasing, and illustrate what we can do to keep waste out of the landfill (Greensboro North Carolina 2023b). The table display suggests how we all can make purchases that avoid excessive packaging, that adds to landfills. At each visit to the market, educators have something new to keep their table interesting and innovative, but also mindful of keeping waste out of the landfill. The recycling educational table often includes a representative from the City of Greensboro education on Protecting Streams and Lakes of NC (Greensboro North Carolina 2023a).

To add to practices that help the environment and reduce waste in landfills, I draw from a student's degrowth experience. He elected to stop using single-use plastic water bottles for the semester and stated, "I wanted to experience what it would require giving up using single use plastic bottles for a twelve-week period." His weekly routine had been to purchase a 24-pack of plastic water bottles. He replaced these bottles with a reusable bottle that he refilled as needed. He tracked the amount of money he saved, the number of bottles he prevented from entering a landfill, the environment, or exported as waste to other countries, and any personal changes. He wrote in his class assignment:

> The environmental benefit from this experience takes priority over the monetary savings of $36.67, in my opinion. Tracking the number of bottles not used over this timeframe recorded I managed to save a total of 246 bottles not going to the landfill. To extrapolate, if I were to impose a one-hundred percent success rate for 250,320 people—the population of my hometown, of not using 24-bottles a week for 12-weeks, it would translate into a savings 71 million bottles not landing in the landfill or elsewhere. And imagine, if the entire population of North Carolina gave up single-use plastic bottles. Such figures suggest an enormous reduction in single-use plastic bottles and landfill waste. (Personal Communication 2022)

Another way to reduce waste is for farmers to come with little to no packing of products, especially those farmers engaged in Community Supported Agriculture (CSAs) projects. A CSA is a direct marketing initiative of small and medium size farmers where customers prepay for a share of the harvest (Henderson 2007). CSA farmers who are using reusable cardboard boxes, cloth bags or other storage containers for their share deliveries. They keep containers to a minimum, minimizing what ends up in the landfill. For example, local farmers, Tracy and Rob created the "Farmers' Choice Box," making all cardboard boxes the same for ease of packing and delivery. Another farmer, Aubry, assisted by his mom, hosted two drop off sites for fresh produce, one at their farm for those who could get there and another in town for those who could walk to pick up their share—each in a brown paper bag with the customer's name and order slip on it.

Coffee vendors at the market are also striving for green consumption and waste reduction by selling coffee in biodegradable cups. For repeat customers they sell a reusable ceramic mug branded with the market's logo. They promote their coffee and reusable mug by offering a one-dollar discount when the coffee is purchased using this mug.

One student who researched coffee for their degrowth assignment argued more could be done with regard to coffee to help the environment and elected to remove coffee from her diet. She reported that over a 12-week period, she saved approximately $63 and did not consume around 1680 fl. Oz. of coffee she would have otherwise been drinking. She only spent $6.74 and consumed 168 fluid ounces in the first few weeks of the course as she weaned herself from coffee. Another student, who said she was addicted to coffee, found out in the first week of recording her consumption that she was spending $50–$60 per week at local take-away places for her caffeine fix. Recording the frequency and amount she spent in only one week shocked this student, which motivated her to reduce her coffee consumption. By the end of their 12-week degrowth experience both students had removed single-use cups and lids for takeout coffee from their routine. The first student who removed coffee from her diet found she went to bed earlier and got up earlier. As she states, "while my total sleep hours did not change, I felt more energized and completed work earlier in the day, providing me more time to do things I enjoy in the afternoons. Overall, I felt like I improved my quality of life as I did things I enjoyed more and spent less of my typical day dedicated to required work" (Personal Communication 2022).

Ultimately, limiting the amount of coffee consumed seemed to resonate with the class, not to mention the need for less coffee beans imported, offsetting the carbon emissions produced by their transport. As a class, we discussed the possibilities of a community-shared experience. It became clearer through these lived experiences that there would be environmental benefits if those who purchased coffee take-out were to reduce their coffee consumption, because less single-use cups, lids, and straws would end up in landfills. Similarly, if those who used single-use prepackaged plastic-cup homebrew at home changed to a reusable cup filter, they too would reduce their contribution to landfills. Such community awareness and behavioral changes can result in fewer single-use items landing in landfills. Drinking less coffee and/or in reusable mugs or biodegradable containers is another degrowth experience and interesting way to keep track of what didn't end up in the landfill by changing individual behavior.[2]

Example 3. Repurposing

Market vendors who repurpose items for sale are not unlike farmers selling value add items of their harvest. For example, a farmer growing cucumbers may also be selling pickles. The market is a place for local artists and crafters to sell upcycled items made from repurposed and recycled goods. One jewelry maker uses seeds to form the basis of her collection. A woodcrafter makes trivets, cutting boards, and bread racks from scraps of wood. Nothing wasted, and old materials are turned into new functional and elegant items. A metalworker enjoys collecting watering cans, teapots, cutlery that are repurposed into garden art and bird feeders.

A nonprofit local business also known as "Reconsidered Goods" emphasizes the use of repurposed material items. Reconsidered Goods takes donated materials from manufacturers and individuals and diverts them from the landfill by offering them to artists, makers, teachers, and kids to create something new. Their mission is to promote environmental awareness, community engagement, and creative educational and artistic expression through repurposing, reconsidering, and reimagining through. During the summer, Reconsider Goods frequently has an interactive table at the Curb Market.

Many customers carry their own recycled bags, cloth bags, straw or wicker baskets to tote their market purchases as they wander the isles visiting farmers and vendors. Bringing their own bags, they minimize the number of single-use plastic bags farmers and vendors give out on market days. Should visitors be short of a bag, there are free bags available for use hanging on a rack at each entrance. Reconsidered Goods created a rack for free bags and a place where community members can donate additional bags. Occasionally, high school students set up a booth at the market with a sewing machine to make free bags from old t-shirts that they recycle into carrying bags for market shoppers. Customers are invited to bring their own T-shirts to be refashioned into shopping bags. Farmers report that because customers come prepared with their own bags or use the available free bags, they purchase fewer single-use plastic bags annually compared to the past.

Example 4. Retooling

The market has become an educational place. Learning about local agriculture, local foods, and seasonality is a quiet and subtle acquisition process. Regulars learn about the seasons for strawberries, blueberries, peaches, and watermelon (listed in the seasonal order). Supporting a local food system allows communities to process and distribute food regionally and sustain its rural heritage. Produce purchased at a local farmer's market has often been picked within the last 24 hours, while supermarket produce has been in transit or cold-stored for days or weeks. This freshness affects the taste of food and its nutritional value. By eating with the seasons, we consume foods that are most abundant and fresh and at their peak taste when they are most nutritious. Supporting local providers fosters responsible land development and sustainable farming.

The Curb Market has become of environmental education hub with monthly displays from the Recycling Center, Climate Interactive and En-Roads Climate Simulation, Water Conservation, Solar Panel companies, Cycling Clubs, and e-bike supporters to name a few groups with a regular presence at the market. They frequently have tables at the market where local representatives inform the public about their organizations' mission, and what they do to get the public involved to help the environment.

The market also includes a lending library. Books on topics like farming, cooking, and personal care are available to the public. Visitors can sign out the books on an honor system and return them when they are finished. Sometimes there are author release events held at the market. There are cooking demonstrations that go together with the donated cookery books. Educational opportunities like tabling, the lending library, and other events transform the community and connect generations.

An indirect way of communal exchange fostered by the market are community members who create their own gardens. Many small farmers sell local transplants for DIY home gardening. John, his wife Sharon, and his sister-in-law sell transplants and offer information and instruction for home gardeners. Sellers who offer transplants make a difference in the community and on dining tables with local and sustainable produce. With rising food prices, including for fresh produce, knowing how to grow one's own fresh produce has allowed some people to eat, sell some at local markets, or share foods with their neighbors. "Shop at the farmers market and eat local" is a core value of many vendors and customers at the market.

Food Justice and Equitable Access

In an ideal world, everyone would have access to fresh produce regardless of economic standing. Like other farmers markets, the Greensboro Farmers Curb Market participates in food security programs. Many farmers accept EBT, SNAP, SFMNP (Senior Farmers Market Nutrition Program) and other voucher programs, which are converted into tokens that are accepted by participating vendors. With supporting grants, tokens available at the Curb Market are doubled in value up to $15. Each token is worth $1 and may be used to purchase fresh produce, fish, meat, eggs, honey, milk, bread, and other food items (https://www.gsofarmersmarket.org/). The Curb Market has an additional token program that is based on "the orange card," a senior voucher program from the Guilford Community Care Network. The orange card token allows for $15 a week to be spent at the market for the items listed above. This program is funded by the Community Foundation of Greater Greensboro and Guilford Adult Health, Inc. (https://www.gsofarmersmarket.org/). During COVID the orange card voucher sponsors allowed a proxy to come to the market to pick up fresh produce for the seniors, something that had not been possible in the past (Andreatta and Hoskins 2022a). All customers can purchase GFCM gift tokens that can be spent like cash at the market with participating vendors. Here the community cooperates to provide fresh farm products for those who need and desire them and may have the least access.

The Curb Market accepts monetary donations from the public to help those in need to purchase at the market. They invite volunteers from the "Farmer Foodshare" program, a nonprofit organization in the City of Greensboro, to set up a donation table to accept food purchased at the market along with cash donations from market shoppers to help others purchase fresh foods from farmers and vendors. They have a table each week at the Curb Market and make their donations to local food pantries. These efforts also help to minimize food waste and prevent fresh farm products from ending up in the landfill or compost pile. Whether it is homegrown or home cooked, keeping it local and out of the landfill is contributing to sustainability and degrowth.

As an urban market, the Curb Market is conveniently located near a bus stop. It has bike racks near the doorways and supports e-bikes. Representatives of e-bike and e-scooter companies and cycling clubs have been to the market to promote non-car urban transportation. Getting around on public transportation or on two-wheels saves gas and limits carbon emissions in a city with three university campuses.

During the semester, one of the students selected to drive less for her 12-week degrowth project. As she found out, she saved money by buying less gas as she walked and cycled more. She shared that the overall experience showed her she did not need to use her car as frequently. She thought if she continued this degrowth path, she would save over 400 dollars annually on gas, 3528 miles (about 5678 km) on car maintenance, and reduce her carbon footprint by 1.16 metric tons. Her new mindful state in limiting car use carried over to time management, specifically making time for the bike ride, walk, or bus trip to get to the needed destination. A year later, she had saved enough money to purchase a new bicycle and no longer owned a car (Personal Communication October 2023).

Mindfulness and Self-Care

Mindfulness is a component of the degrowth movement. Philosophically degrowth is a lifestyle coupled with a mindfulness or an awareness about one's connections to people, family, nature, or the arts. Some individuals might want to work less or be involved in workshare to have more family time, engage with nature, or have time for other activities, such as the arts, sports, hobbies, and leisure. This mindfulness aligns with believing money is not everything and that social connections are crucial to one's quality of life. It is conscientious decision making for actions, behaviors (purchases), interactions with others, and the Earth. Mindfulness is about taking care of oneself emotionally, mentally, and physically. The Curb Market fulfills these needs for many regulars. People come to the Market to socialize, visit, and support their community, local farmers and vendors.

Carol, a vendor and owner of Mimi's Soaps, a home-based business of artisan soaps best illustrates issues of self-care. Since 2001, she has been making handcrafted care and cleaning products like soaps, facial creams, body lotions, lip gloss, laundry detergent, and candles from natural ingredients. Many of her customers are partial to her products, particularly shampoo and bar soap, due to the fragrance options and limited packaging she uses. Others purchase her laundry detergent because she takes back and reuses the containers.

Conclusions

In this chapter, I highlighted how market managers, farmers, vendors, and visitors are transforming the GFCM by reusing and repurposing items, limiting the use of single-use plastic items (bags, cartons, clamshells, and jugs) and single-use glass, and cups for coffee and tea. As a regional market, the GFMC limits carbon emissions because fresh farm and artisan products travel only short distances. Collectively farmers, vendors, and customers combat climate change and keep items out of the landfill. Together, community engagement and thoughtful individual attempts, as illustrated by the student projects, can produce a small greening effect. Buying local, limiting packaging and plastics, repurposing or reusing items and learning more about how to create a more sustainable lifestyle, a community of market supporters strive for sustainability at the Greensboro Farmers Curb Market.

Students who participated in the degrowth project recognized their role in making an environmental difference in their community. I introduced student projects that documented behavior changes to show what is possible and how more people could follow their actions. Turning these behavior changes into everyday routines early in the semester, these practices became normalized and ordinary by the end of the semester. We learned that each person can change one practice and contribute to change. If more people adopted similar practices to reduce waste going to landfills, reduced overall consumption, connected more with nature to value and protect it, and encouraged more people to join such efforts, we can all help to clean up the environment. Degrowth might not be a roadmap for everyone, but it can help to encourage many individuals to adopt concrete practices that help the environment.

Notes

1. All individuals' names are used with their permission and are their farm or business names. All interviews and observations were conducted between March and October 2022 at the Curbside Market, and from an anthropology

class I taught at the University of North Carolina at Greensboro in Spring 2022.

2. There are plenty of online examples of biodegradable containers However, some have elected to use mason jars, even for hot beverages. They wrap the bottom and mid-point of the jar with yarn (nonacrylic) to protect their skin from the beverage's heat.

References

Andreatta, Susan. 2021. Moving from Reading to Dialogue to Action: Teaching Degrowth in Anthropology Courses. *Teaching and Learning Anthropology Journal* 4 (2): 49–68.

Andreatta, Susan, and Mia Hoskins. 2022a. Covid-19 Responses and Pivots in the Piedmont Region of North Carolina: A Call for Policy Reforms for Small Farmers. *Human Organization* 81 (4): 316–326.

Andreatta, Susan, and Mia Hoskins. 2022b. Supporting a Local Agro-Food System in Covid-19 Times. *Anthropology and Ethnology Open Access Journal* 5 (2): 000177.

Bonnisseau, N. 2022. Is It Too Late to Save Our Planet? https://plana.earth/academy/is-it-too-late-for-our-planet. Accessed 11 September 2023.

D'Alisa, Giacomo, Federico Demaria, and Giorgos Kallis. 2015. *Degrowth. A Vocabulary for a New Era.* New York: Routledge.

Godin, L., and J. Langlois. 2021. Care, Gender, and Change in the Study of Sustainable Consumption: A Critical Review of the Literature. *Frontiers in Sustainability* 2:1–11.

Greensboro Farmers Curb Market. 2024. https://www.gsofarmersmarket.org/. Accessed 13 January 2004.

Greensboro North Carolina. 2023a. Protecting Streams and Lakes. https://www.greensboro-nc.gov/departments/water-resources/get-involved/protecting-streams-and-lakes. Accessed 11 September 2023.

Greensboro North Carolina. 2023b. Recycle. https://www.greensboro-nc.gov/departments/field-operations/recycle. Accessed 11 September 2023.

Greensboro North Carolina – Government. 2023. City of Greensboro Growth and Development Trends. January. https://www.greensboro-nc.gov/home/showpublisheddocument/54749/638092073631970000.

Henderson, Elizabeth. 2007. *Sharing the Harvests: A Citizen's Guide to Community Supported Agriculture.* 2nd ed. Vermont: Chelsea Green Publishing.

Heuer, J., G. Lehigh, K. Diaz-Serrano, N. Romero-Daza, and D. Himmelgreen. 2020. Food Access in the Time of Covid-19: Reflections on the United States Food System During a Pandemic. *Practicing Anthropology* 42 (4): 13–16.

Kallis, Giorgos. 2015. The Degrowth Alternative. Great Transition Initiative. February. https://greattransition.org/publication/the-degrowth-alternative.

Kallis, Giorgos, Vasilis Kostakis, Steffen Lange, Barbara Muraca, Susan Paulson, and Matthias Schmelzer. 2018. Research on Degrowth. *Annual Review of Environment and Resources* 43 (1): 291–316.

Latouche, Serge. 2004. Degrowth Economics: Why Less Should be so Much More. Le Monde Diplomatique. November. https://www.jussemper.org/Resources/Economic%20Data/Resources/Degrowth%20economics,%20by%20Serge%20Latouche.pdf.

Latouche, Serge. 2009. Farewell to Growth. Polity Press. Cambridge UK.

Muraca, Barbara. 2013. Décroissance: A Project for a Radical Transformation of Society. *Environmental Values* 22 (2): 147–169. https://doi.org/10.3197/096327113X13581561725112.

Paulson, Susan. 2017. Degrowth: Culture, Power and Change. *Journal of Political Ecology* 24:425–448. https://doi.org/10.2458/V24I1.20882.

Paulson, Susan. 2020. Degrowth and Feminism Ally to Forge Care-Full Paths Beyond Pandemic. *Interface* 12 (1): 232–246.

Saitō, Kōhei. 2020. *Slow Down. The Degrowth Manifesto.* Translated by Brian Bergstrom. New York: Astra House.

United Nations. 2021. COP26: Together for Our Planet. https://www.un.org/en/climatechange/cop26. Accessed 31 December 2022.

United Nations. 2022. UN Climate Report: It's "Now or Never" to Limit Global Warming to 1.5 Degrees. https://news.un.org/en/story/2022/04/1115452. Accessed 31 December 2022.

Wagner, Travis P. 2017. Reducing Single-Use Plastic Shopping Bags in the USA. *Waste Management* 70:3–12. https://doi.org/10.1016/j.wasman.2017.09.003.

CHAPTER 5

"These Systems Will Provide Their Own Lessons": University Reuse Programs as Sites of Material Care and Educational Praxis

Angela D. Storey and Henrietta K. Ransdell

Introduction

On a breezy Tuesday in late 2021, author Ransdell walked into the Red Barn, an iconic student space on the University of Louisville's campus. A friend and fellow student sat at the front desk welcoming folks into the Free Store's monthly "Free Sale," where anyone could donate gently used items and others can take them for reuse. The Free Store had closed during the construction of a new dorm, but students had organized these pop-up events in response. The store had recently found a new, albeit small physical home, in a storage closet in the Student Activities Center, but the popular monthly events continued.

At the Free Sale, tables and clothes racks were jammed with colorful shirts, pants, shoes, and more. Pleased at being early, Ransdell found two new-to-her sweaters and perused a tantalizing pile of second-hand board

A. D. Storey (✉) • H. K. Ransdell
Department of Anthropology, University of Louisville, Louisville, KY, USA
e-mail: Angela.Storey@Louisville.edu

P. Kuppinger (ed.), *Second-Hand Cultures and Economies of Reuse, Repair, Sharing, and Care*,
https://doi.org/10.1007/978-3-031-99875-1_5

games. From the hesitant meander and glances of passersby, she saw in action something we had learned by observing previous Free Sales: it can be hard to convince people that donated items are truly for anyone, and that they can take anything without needing to reciprocate by donating items themselves. While it requires some convincing at times, people do take items, and Free Sales can have hundreds of visitors. The message of the sales is clear: diverting waste means putting things in the hands of someone who will give them new life, with no need to prove you deserve them.

We live in an era defined by capitalist relations of consumption, waste, and toxicity—a time called the "wastocene" (Armiero 2021) by some scholars. Surrounded by waste and by things headed towards a ready disposal, our relationships with material objects can be understood anthropologically by exploring the cultural constitution and symbolism of waste as a category (Douglas 2003 [1966]) as well as by examining what patterns of material use and discard can tell us about relations of life (Isenhour and Reno 2019).

University campuses are spaces with intense material use and, frequently, the high generation of waste. Academic analyses of campus waste generation have often focused on the division into manageable streams for disposal and the possibilities of recycling programs to capture reusable resources (cf. Armijo de Vega et al. 2008; Baldwin and Dripps 2012; Adeniran et al. 2017). These institutional practices, though, can go hand in hand with smaller programs that center specific forms of material reuse.

This chapter explores spaces of reuse at one U.S. university to examine what they mean for students engaged in their creation and maintenance. We define reuse broadly, as ways in which waste reduction is also generative of another material use, whether that be a direct reuse or a transformation of the materials into something new, and we emphasize programs in which the recirculation of waste intersects directly with student experiences. Drawing from a three-year, participatory study of student engagement in the University of Louisville's Free Store, Food Recovery Network, campus garden, and composting programs, we explore what spaces of reuse mean for students as material care becomes a form of educational praxis (Mahon et al. 2020). We focus on interviews conducted with undergraduate and graduate student sustainability workers and participant observations conducted at campus reuse-focused events held between 2021 and 2023 to examine how students leading these programs talk about their experiences and the impact on their lives. As students articulate the power of these spaces for learning, they highlight the importance of relational experiences. Collaborative work centered on reuse becomes a way for students to express care for each other and to craft the kinds of socio-material worlds they want to exist.

Repair and Care

A growing interdisciplinary literature on repair and care explores reuse practices as relational acts. Scholars emphasize how material objects and practices make tangible social connections and forms of caring labor, challenging us to not only reimagine waste but also the social practices that surround it.

Work in discard studies suggests we often consider that "waste is the opposite of commitment, care and sustainability" (Martínez 2017, 347), the result of processes that have "failed" or are "unfinished" (ibid.). What, though, counts as waste? A deeply social question, waste remains not only that which symbolically defies cultural domains, as Douglas contended (2003 [1966]), but also that which we define materially as superfluous to needs. Practices of reuse challenge waste's "failed" or useless state, evoking instead the possibilities for material practices of reclamation. Indeed, "reuse involves deliberate acts of revaluation and care which recall and build upon embedded meaning, affect, social histories and the properties of materials" (Isenhour and Reno 2019, 1).

Practices of reuse can be overtly political acts, meant to defy capitalist drives towards endless consumption and profit. They can also be forms of making-do in a world in which resources are inequitably distributed. Often, reuse programs bring together these two goals, challenging a "throw-away" cultural norm while also channeling useful materials into the hands of those who can use them. While this pairing can highlight reuse as an ethical stance, it also demonstrates the flexibility of reuse as a category of action, one in which a single act could be interpreted as either one of resistance or of necessity. What draws these ends together is an understanding of reuse as a form of care practice, one in which material objects manifest social relations. As Isenhour and Reno write, "Caring for objects…could be said to involve both attention to a thing's material qualities, as well as a concern for why and for whom they matter" (2019, 3). To articulate care as central to reuse highlights what it means to evoke more than affect but also to emphasize what Puig de la Bellasca calls "a notion of material doing… an ethically and politically charged *practice*"

(2011, 90, italics in original). Care is thus a commitment, a practice, an action; something with both tangibility and agency in it.

In this chapter, we explore what it means for students to evoke spaces of reuse as forms of care, ones in which connection and meaning emerge from not only processes of learning but, critically, in kinds of educational praxis that bring together learning and doing.

Background

As environmental issues received increasing attention and concern in the 1980s, colleges and universities started to adopt sustainable practices (Emanuel and Adams 2011). Thinking of sustainability as "meeting the needs of the present generation without compromising the ability of future generations to meet their needs" (Brundtland Report 1987), many such efforts have been aimed at engaging students and on integrating sustainability into the curriculum (Cogut et al. 2019). Indeed, student participation and leadership have been integral to the success of campus programs (Helferty and Clarke 2009), which then facilitates community building for students and other university members (Crutchfield et al. 2020). Sustainability programs are also linked to enhanced awareness of environmental topics; one study found that the presence of a composting program on campus was correlated with higher student participation, more knowledge, and related behavioral changes (Waliczek et al. 2016). Engaging students in hands-on sustainability work can also be core to educational outcomes, creating living laboratories for student learning (Evans et al. 2015).

At the University of Louisville (UofL), the majority of sustainability programs are coordinated through the campus-wide Sustainability Council and via the labor of one full-time sustainability staff member and part-time work from a faculty member. Begun in 2008, the council's work includes weekly programming throughout the academic year, ranging from hands-on workshops in the campus gardens to lectures about green burial practices, beekeeping tutorials, and community fairs. With limited staff, significant programming runs through the work of paid annual student interns. Other, related programs of reuse are run in conjunction with the student government and student affairs, which also employs student workers. For this chapter, we explore four programs: the Garden Commons, the composting program, the Free Store, and the Food Recovery Network.

The UofL Chapter of the Food Recovery Network (FRN) is a group of student volunteers, coordinated by a paid student intern, who collect edible leftover foods from campus dining locations and divert them to the Cardinal Cupboard on-campus food pantry or to other nonprofits in Louisville. Students train one another on food safety practices and the challenges of food insecurity before participating in one-hour "runs" to transport leftovers from food service locations as they close up for the day. Through this practice, FRN has recovered over 10,500 lbs. of food from UofL campus dining locations (Food Recovery Network 2023) and helped draw attention locally to national problems of food insecurity on campus (Henry 2017; Cady and White 2018). As FRN is closely tied into the Cardinal Cupboard, we speak about them here in tandem.

At the campus Free Store, gently used items are collected and offered for reuse. From clothes and household materials to cooking-ware and books, student volunteers give new life to materials donated by the UofL community. Students, faculty, staff, and community members are welcomed to donate and "shop" at pop-up free sale events on campus, as described in our opening vignette, or in the Free Store's storage closet. The weight of items is tracked by the student interns who coordinate the Free Store to acknowledge pounds diverted from the landfill.

The Garden Commons consists of two small student-run campus gardens that provide free harvests of organic produce. Students and community members alike are invited to weekly workdays and monthly workshops throughout the academic year, coordinated by student interns, where attendees can gain the skills to grow their own food and learn about sustainable food systems. The free-access model allows anyone to use the gardens as a resource for food or learning at any time, and the gardens link to other campus sustainability programs as well, donating produce to the food pantry and utilizing compost generated on site.

The UofL Community Composting project generates around 4000 pounds of compost per week from pre-consumer food waste (UofL Sustainability 2023). Student interns collect organic food scraps and coffee grounds from campus dining facilities and community business partners, then lead weekly compost-turnings to teach volunteers how to generate their own nutrient-rich soil. Community members are encouraged to take finished compost, in addition to it being donated to the UofL Garden Commons and other local urban gardens.

Methods

This project utilized participatory and collaborative ethnographic methods (Lassiter 2005) to develop and conduct research guided by the workers and volunteers within campus sustainability initiatives. We utilized participant observation and semistructured interviews, and data was collected beginning in spring 2021 and finishing in summer 2023. A total of 22 semistructured interviews were completed, including individuals in two categories. First, 17 interviews were completed with UofL students who were employed and/or otherwise compensated to conduct sustainability-related work on campus. This included undergraduate, graduate, and professional students who held any type of formal, named position that included compensation (financial compensation in the form of hourly pay, stipend, work study, fellowship, or scholarship; or other forms of compensation, such as credit hours for an internship); most were Sustainability Council interns paid hourly. Second, five interviews were completed with staff and faculty engaged in coordinating Sustainability-related programming and, especially, in managing and interacting with students involved in sustainability work on campus.

Participant Observation was conducted within a variety of spaces and events related to campus sustainability initiatives, including volunteering days at the UofL composting site, Garden Commons, and Food Recovery Network; in-person events such as maple tapping and foraging workshops; and volunteer times within sustainability programs, such as regular FRN food pick-ups. A total of 41 fieldnotes were completed for 34 different events observed between 2021 and 2023.

The collaborative and participatory nature of this project was central to its design and execution, with a total of 14 students engaged in research design, data collection, analysis, and writing alongside the faculty lead (Storey). Of these, two graduate students were central to the coordination of the project, and 13 undergraduate students were involved for between one semester and one year (with one student, author Ransdell, involved as an undergraduate and graduate student). Nearly all of the student researchers were employed by sustainability initiatives on campus, completing degrees in Sustainability, and/or active within the programs examined in the project, and thus the research relied upon and intentionally built from the active engagement of students in these programs. For two years of the project, author Storey served as chair of the university Sustainability Council and thus also engaged in this research as a program participant.

We position our work as an engaged project that sought to build deeper understanding from existing knowledge. This research began during the COVID-19 pandemic and was designed to be conducted outside, online, or in socially distanced ways.

Data and Analysis

In this section, we explore the ways in which students spoke and wrote about their experiences working in and inhabiting spaces of reuse on campus. We begin with a discussion of how programs offered informal spaces for learning and then bridge into a discussion of how those experiences of learning reflected, created, and compounded connection, meaning, and agency for students.

Many students began their connection to reuse-centered programs through formal, institutionally coordinated avenues. This included classes, internships, degree programs, and events coordinated by the Sustainability Council as well as related programs. Faculty promote these programs as resources for the students, including food available from FRN, the Garden Commons, and campus food pantry, and they even utilize the sites for class-related volunteerism. As Chris, a student worker in the compost program explained, "we'll have people from ...the biology department (come) to get either samples, or worm populations for their classes, or to come pick up compost for like the pollinator garden" (pseudonyms are utilized for research participants).

Despite the connections that do exist to formal educational spaces on campus, most students spoke about reuse sites as spaces for informal, experiential learning and the development of an out-of-class praxis that often relied upon learning from peers. For example, in observations conducted at the campus composting site, located in a work yard at the north end of campus that is unknown and unvisited by many campus members, student researchers were consistently impressed by how much they learned through shared labor. The compost site is an open area, centered on piles of food slowly becoming compost, ringed by moats dug into the wood chip subsurface. Around that sit a variety of related experiments in sustainable living: beehives, compost tea-making machines, aquaponics projects, and greenhouses among them. It is a place that feels ripe with possibility for building alternative lives—and also ripe with the feeling of how that is made from the (sometimes smelly) processes of reuse.

For student researchers who also worked as compost interns, their fieldnotes highlighted the ways in which their own knowledge, built through labor, connected them with peers. Marisa, a student researcher and compost intern, wrote in these lightly edited fieldnotes in April 2022 about the experience of becoming the knowledgeable peer on site as she interacted with her supervisor, another intern, and a volunteer, and recollected a former student mentor at the site:

> Without even asking my supervisor, I know what to do. I say to another intern and a volunteer "hey, how's it going?" They both smiled and said that they were good. The new volunteer told me her name and that she was here to fulfill volunteer hours for her sorority. My fellow intern started working here last week, so she's very new. The volunteer told me that this was her first time at the composting site. She was using a shovel; as the person with the most experience out of us three, I suggested using a pitchfork instead because it picks up more material (less time consuming) and is physically easier to use. The first time I ever turned a pile, I also tried to use a shovel. I remember the former intern who told me to use a pitchfork instead. Oh, how the tables have turned.

Here Marisa moves from being a learner to a teacher, sharing tips for improved practice with peers and appreciating the cyclical nature of this space's learning experience, as she shares practical knowledge onwards.

Similar processes took place at the garden sites on campus, one location of which is tucked in a protected and chaotically growing green area behind a dorm, a parking lot, and an office building. A space filled to the brim with life and experiments in sustainability, several edible native plants grow here—paw-paws, maypops, mulberries, and serviceberries among them—along with more traditional garden vegetable and some fruit trees, including a tree claimed to be grafted from the one that dropped its apple on Isaac Newton's head. Author Ransdell wrote about her experience of peer connections in the following edited fieldnotes from fall 2021 about a campus garden work party here:

> Forty minutes had passed, bending up and down, up and down, clearing away the spider-web of maypop vines that had consumed the garden pathways. It was amazing what ten people could accomplish in an hour; the whole garden was transformed. I showed the plastic bags full of overripe maypops that a friend and I had collected in the process; we had hundreds of them. Minutes ago, my friend Elena had just taught me about the bitter

> taste of their flowers and how to crack and eat the round fruits. The nostalgia of eating figs right off the branch at my childhood home flooded over me, so I had to show everyone else too. One volunteer was hesitant, never having encountered maypops before. Elena pulled one open to show her, and the volunteer let out an 'ewww' as the odd texture inside the fruit spilled out. Everyone tried one and were surprisingly pleased with their sweet tang. I was relieved, as we had an awfully big harvest to send home.

Here the combination of the work and discussions about it produced not only new knowledge of a native edible fruit but also connections between peers.

As experiential learning is increasingly touted as a high-impact learning practice (Kuh 2008) with significant outcomes for students across disciplines, spaces of interaction that move from learning into connection offer ways for students to see themselves as both learners and teachers. Students bring into these informal spaces their own funds of knowledge (Vélez-Ibáñez and Greenberg 2000), sets of cultural resources and diverse lived experiences, which might not fit into classroom spaces but here are shared and appreciated. Approaches that embrace a diversity of knowledges and skills, and which highlight ethical stances and engaged processes, emerge as forms of educational praxis (Mahon et al. 2020). Here, praxis emphasizes situated, concrete, and responsive experiences of *doing* that center forms of justice and change-making in the world (ibid.).

When asked why they continue the work they do, students responded almost unanimously with the sense of community and the relationships they had developed. Chris, the composting intern and a Sustainability major, saw his work as "an enjoyable way to spend time and to meet people and to connect". He emphasized that, while compost is the product that brings many to the site, it's the relationships that keep people there. Chris was a senior at the time and told us that he was still in contact with some people he had met composting as a first-year student. He explained what the diversity of volunteers meant, saying:

> It's not uncommon to have like at least ten people out there on a Sunday, all, all from different sections of Louisville, doing different things...you're gonna be able to connect with somebody that will help you with something. ...I was thinking about it the other day actually...whether I would do like this exact thing if I wasn't being paid to do it, and I think I would. I think like my favorite part about it is the fact that I get to interact with community members.

Similarly, Mia, a former garden intern, highlighted the process of building community as one of the biggest impacts, saying that she "didn't have a sense of community on UofL's campus until I became a garden intern". The internship began a snowball effect: meeting friends in garden workshops led to recognizing them in class or outside her major, and then joining other organizations on campus that they told her about.

Many students highlighted how reuse programs brought people together, framing it as a critical part of their involvement. Beginning in an electrical storage closet in the student activities center, the Cardinal Cupboard and Food Recovery Network were organized by students to collect leftover bagels and pastries on campus and redistribute them to students and community members. Around 4 PM every weekday, you could spot a student wheeling around a red fabric cart scattered with prepackaged salads, bagels, muffins, and cake-pops, often stopped by their peers for free samples. Cheyenne, a student researcher on this project, highlighted such links in her fieldnotes as she described an FRN pick-up in January 2023, writing:

> We begin our journey to pick up the leftover bagels from the day, the 5-minute walk consisted of (the FRN intern) teaching me more about the cupboard and answering questions I had… We also had more personal conversations such as our majors and how we are connected with food. (She) is a current nursing major which surprised me, as every other sustainability intern that I have met is either in the major or something very close to it. She decided to be the FRN intern because she used the cupboard often her first years on campus and felt connected to it.

Rachel, a food-pantry worker and a political science and economics double major, recounted how her motivations for participating in sustainability work transformed from material to interpersonal. With food donations and collaborations with local grocers, the food pantry outgrew the supply closet. Nestled between other amenities like athletics ticketing and student affairs, the Cupboard now hosts silver-lined shelves with staple foods, produce, and even hygiene products. Rachel spoke of her unexpected involvement as a core student worker:

> It was a service opportunity, and I never really saw it going this far. Both in terms of the Cupboard's growth itself as well as my involvement with it. That was really how I got started, just word of mouth, and then I kinda got roped in, I had some conversations with people about how it would be really

> cool if we also offered these opportunities to people, or let them know about these resources, and things have just expanded.

Being connected to others not only showed Rachel how she could contribute more herself, but also demonstrated the power of such linkages for keeping people involved. The food pantry's connections in particular span across Louisville, including community members, business partners, and other service initiatives.

Author Ransdell, who was one of the founders of the food pantry and knew how closely the space linked to city projects for food security and other programs on campus, observed this resonance during a campus tour in the fall of 2021. She wrote:

> The space certainly feels like a hub with many spokes connecting to different groups and places throughout Louisville; I don't know if I have this impression due to my intimate knowledge of the space or from the environment itself, but I cannot shake the sense of community it brings to me every time I am there.

While students find possibilities of connection throughout their university experiences, something specific seems to happen within the labor at sites of reuse—experience that not only shapes individuals but also shapes people in relation to each other and the materials around them. As Isenhour and Reno write, "using and reusing offer affordances for becoming kinds of people and for shaping social relations far beyond consumer identities. Since waste practices extend social relations either way, reuse in however humble a form gives consumers more of a role in shaping the fate of discards, and thus shaping themselves" (2019, 4). As students highlight the small spaces of connection and discussion found while turning a pile of compost, picking up bagels at the end of the day, or gathering maypops in the garden, they seek out and solidify elements of themselves that emerge in such interactions—elements that are called forth by the topic of their work and the tangibility of their productive endeavors. Reuse spaces depend upon significant and continuous labor for the processing of materials, which, in this case, offers opportunities for students to take on leadership roles and build skills beyond the classroom.

These campus spaces create avenues for care that connect students to the tangible and intangible: environment, (non)waste, and the people laboring in between. While coordinated by separate interns, the reuse programs on campus do connect and, in ways, create a circle of support. Rachel, the food pantry intern, pointed out how the Garden Commons utilizes compost from the campus site, then gives produce to the food pantry, who in turn brings organic waste back to the site.

These small spaces of connection layered further meaning onto the work itself, which students often spoke about as personally very important. Marisa, the composting intern and Sustainability major who shared her pitchfork knowledge with peers, wrote about this kind of meaning on a wintery day in February 2022:

> As the sun goes down, the temperature is dropping. I have water inside my shoes and gloves, it's muddy, my clothes are very dirty, I'm freezing, and ready to go home. My coworker is saying "we're almost done!" Her encouragement is helpful. Once again, I start thinking about how satisfying it feels to be a part of this process, regardless of the obstacles (such as working in cold and freezing weather). Sometimes I just look at it like work, but as I'm looking around this site today, I can see so much evidence of hard work and dedication. I am reminded of the amount of effort it takes to maintain this composting operation. My coworker and I finally finished cleaning the buckets. While we were turning off the faucet and unscrewing the water hose, I asked her if she also experiences the feeling of satisfaction by working here. She said yes, because it makes her feel like she is doing something for the community, environment, and herself.

The feeling of being part of something larger than yourself, and in doing work that felt like it made a tangible impact, was critical to why students enjoyed and continued this work—even when it meant freezing cold compost days, unsuccessful FRN runs, or needing to convince peers that goods at the Free Sale really were free.

Mia pointed out a kind of cycle of care in discussing how she now teaches the skills once taught to her by a friend and previous intern at the garden, just as Marisa spoke about learning composting skills from peers and then passing them on to new volunteers. Teaching skills to others offers a way to showcase small moments of agency and empowerment, especially in the context of hegemonic processes that expect consumption and waste. As Mia said:

> I feel like the more skills you have, the more power you have as an individual, especially when it comes to anything that (is) kind of opposing mainstream in terms of being able to grow your own food and being able to take

your food to composting sites rather than putting it in a landfill, or taking your clothes to a free store rather than being able to throw them away.

The productive reiteration of tasks passes on knowledge and builds agency, demonstrating why these connections to community and peers resonate across responses: they are founded upon care.

One of the most poignant gifts students generated amongst themselves was their collective feeling of agency. Eric, a junior Geography and Sustainability major who worked as an intern for the zero-waste initiatives on campus, spoke about the role of sustainability work on campus, saying of shared concerns and worries of limited action, "I think it gives students a platform to…do something about it." He continued on to say that "students are the ones that actually push for the changes." Senior Engineering major Mason, who coordinated the Food Recovery Network, likewise highlighted how many sustainability programs existed because of the labor and initiative of students. He said, "It's a student led endeavor .. it shows that like our own students care about helping each other. It's not just a top-down thing." Rachel was excited to say of the food pantry, "I worked in an organization that helped the university to change its practices," with students committing over 1000 hours of community service to serving more than 8000 individuals. She saw how the effort students put into these programs shifted the culture of the institution at large, saying: "I hope that's the biggest impact the sustainability organizations across campus have is really just to change that dialogue, to open the university to new ideas, and to help them to see that…this is a way that we could become a big change maker." Here, the labor of reuse is a way to showcase the import of students to the university's functioning and even its ethical foundations within the community, as students run programs, support work, and also call for wider change. Like Rachel, Mason, and Eric, many students saw the work they did in reuse programs as work that would likely not exist without their labor and dedication, which reinforced their own role in making change and in crafting spaces that prefigure wider worlds they would like to see.

Students also reflected upon their campus engagement work as an antidote to classroom experiences. Chris told us that in thinking about the scale of global environmental concerns, "you just feel powerless, especially when you're sitting in a classroom just reading about it." Asked about how this linked to work at the compost site on campus, he replied:

> My favorite part about it is the fact that I get to interact with community members and have these conversations like around these topics with them, and I don't know, get points of view that I'm not necessarily getting in a classroom, from a book about it… and also to be able to like offer something in return to them that's tangible.

Likewise, Mia spoke about the gardens and composting programs as "pretty unique experiences (in which) people learn some cool skills that you wouldn't necessarily learn in a classroom setting." She continued on to talk about how such spaces allow students to build "knowledge about things that maybe aren't typically accessible to students."

Across these programs and the work of students, collaborative labor emerges as a form of care. Students garner pride and satisfaction from laboring alongside one another and seeing the change that work can make. They also open conversations on sustainability to their community and change the narrative of what an ethic of care can look like in higher education. Students saw and spoke about their work as a two-way street, where the institution shapes them, and they shape it right back. Dominic, an Economics major who had held an internship with the Sustainability Council and also was Director of Green Initiatives for student government, spoke about being a student worker as being able to see "behind the veil" of the institution, able to see more of how the university works and also to shape it.

The labor within reuse spaces highlights how material care becomes social care. These spaces of experiential learning are ones managed, shaped, and created by students themselves—and are spaces defined by kinds of knowledge and labor that are often not relevant in academic settings. Indeed, the narratives that emerged from interviews and participant observations are ones about work, about laborious interactions with materials that students are working to keep from the waste stream. It is the labor of turning smelly compost, of picking up end-of-day sandwiches, of sorting piles of donated clothing—of care for things that becomes care for each other. Mia described how these forms of labor entwine as she spoke about the garden, saying: "that type of work, it's really meditative, it brings me a lot of joy, and it also is really joyful that it's like engaging in community and also collective work, being able to grow food with people and then for people that's for sharing is a really cool aspect of community that I enjoy a lot, too."

These experiences also do things *to* the students who manage and participate in them, some of them are quite unexpected outcomes but also impactful ones. Most of the student workers were hired through internships, and many reflected upon the impact in terms of expanding their resume and exploring career options. Dominic told us of the usefulness of his different sustainability roles: "you make these connections and it kind of helps you like know what you want to do after you graduate or with your life in general. I don't know what I'd be doing if I hadn't joined (student government), but probably not the exact same thing that I'm doing now." He continued on, considering the role of campus opportunities as "little doors" that allowed students to explore long-term possibilities: "the(se) little doors are very helpful for students … it's very specific in helping students kind of find where they fit and like what they want to do with their life."

Other students spoke about how these experiences shaped their life trajectory in unexpected ways. Mia, the former garden intern who continued volunteering at the compost site after graduation, spoke about how the internship provided not only skills but also a set of alternative career options and interests that had not come from her degrees. Talking about the internship, she said: "that was my first ever exposure to growing food … it was very empowering because it taught me an essential skill." She continued, "I feel like it kind of just like propelled—if that's the right word—(me) down this... new path of interests that I found of food-growing mixed with food justice and sustainability." For both Dominic and Mia, the internships and formal positions offered clear avenues of educational praxis through which to build their interests and skills, and to find new spaces outside their disciplines for engagement and, potentially, future work.

As students make connections to each other through their work in reuse sites, they also see the wider importance of reuse programs in terms of removing items from waste streams and in meeting immediate needs. As Claire, who ran the Free Store, explained:

> So many people have come up to me, have come up to my volunteers and told me "thank you so much for doing this, I needed this specific item." …one example was a lady needed a laundry machine and there was a portable washing machine that had just gotten donated. She got that and she was like, "You know, thank you so much, I didn't have one in my apartment. I

> didn't know what I was gonna do, but... (now) I don't have to worry about what's gonna happen next."

Rachel also spoke about this in relation to the food pantry and FRN, saying: "I think it is a very personal measure of success just to see somebody's face at the end of their shopping...to hear people say...this will get me to this end of the week...hear that we've made some impact in someone's life because the numbers don't reflect (the full impact)... How do you quantify five people suffering from food insecurity? Like, it's much deeper..."

Meaningful interactions like these shifted the students' perceptions on sustainability and how they recognized their roles and impacts. Free Store Manager and Music major Claire recognized the possible ripple effects of student action even beyond the university, saying: "I think having (the free store) on campus is a good way to start that nationwide, hopefully in other communities as well." She spoke about how much she had learned about sustainability through the internship, and also her growing expertise on the problems of fast fashion and consumption. Most students spoke about how their perspectives had grown during their work. Eric, the zero-waste intern, told us that he "came into this (role) thinking that all we need to do is just fix climate change and bring down emissions, but...it's a much more complicated problem than I thought it was ever going to be. Realizing that sustainability really impacts your entire life, not just one part of your life."

Through tangible work towards sustainability, campus is made into a place resonant socially and also physically (M'Gonigle and Starke 2006). The labor of students is central to these outcomes and to making these programs ones in which care emerges. As Puig de la Bellasca writes, "Caring is more than an affective-ethical state: it involves material engagement in labours to sustain interdependent worlds" (2012, 198). Here, learning and doing combine to become educational praxis.

Conclusion: Reuse Does Things

Reuse programs make change—for the students, the campus, and potentially beyond that. Within them, they hold possibilities for creating and nurturing relations and practices of care that depend upon the labor, agency, and attention of students, and which bring to the fore forms of knowledge and praxis that are often not valorized within classroom spaces. As one of the sustainability staff said of the reuse programs, "These systems will provide their own lessons." Through their work, students craft spaces that reflect the world and culture they want to see—spaces that highlight reuse as an ethical stance and a flexible set of repeated practices, a form of care that is a kind of "material doing" (Puig de la Bellasca 2011, 90).

Here the material and relational converge, as forms of connection and feelings of belonging emerge from the labor of making things useable in new ways. This happened often in small ways: a composting volunteer recounted the excitement they felt upon noticing that someone had shoveled the compost pile earlier that day to make her job easier. The satisfaction student workers share from being a part of something bigger than themselves represents a culture of reuse that's not simply transactional, but regenerative. Whether shoveling compost, re-homing washing machines, or making friends at the garden, reuse spaces on campus provide opportunities for students to make spaces producing change, both for the community and for themselves.

References

Adeniran, A. E., A. T. Nubi, and A. O. Adelepo. 2017. Solid Waste Generation and Characterization in the University of Lagos for a Sustainable Waste Management. *Waste Management* 67:3–10. https://doi.org/10.1016/j.wasman.2017.05.002.

Armiero, Marco. 2021. The Case for the Wastocene. *Environmental History* 26:425–430.

Armijo de Vega, Carolina, Sara Ojeda Benítez, and Ma. Elizabeth Ramírez Barreto. 2008. Solid Waste Characterization and Recycling Potential for a University Campus. *Waste Management* 28:S21–S26.

Baldwin, Erika, and Weston Dripps. 2012. Spatial Characterization and Analysis of the Campus Residential Waste Stream at a Small Private Liberal Arts Institution. *Resources, Conservation and Recycling* 65:107–115. https://doi.org/10.1016/j.resconrec.2012.06.002.

Brundtland, G. H. 1987. *Our Common Future: Report of the World Commission on Environment and Development.* New York, NY: United Nations General Assembly.

Cady, C., and C. C. White. 2018. Food Pantries on Campus to Address Student Hunger. *New Directions for Community Colleges* 2018 (184): 73–82. https://doi.org/10.1002/cc.20329.

Cogut, G., N. J. Webster, R. W. Marans, and J. Callewaert. 2019. Links Between Sustainability-Related Awareness and Behavior. *International Journal of*

Sustainability in Higher Education 20 (7): 1240–1257. https://doi.org/10.1108/IJSHE-09-2018-0161.

Crutchfield, Rashida M., Andrea Carpena, Tahirah N.,McCloyn, and Jennifer Maguire. 2020. The Starving Student Narrative: How Normalizing Deprivation Reinforces Basic Need Insecurity in Higher Education. *Families in Society: The Journal of Contemporary Social Services* 101(3): 409–421. https://doi.org/10.1177/1044389419889525

Douglas, Mary. 2003 [1966]. *Purity and Danger: An Analysis of Concepts of Pollution and Taboo.* New York, NY: Routledge.

Emanuel, R., and J. N. Adams. 2011. College Students' Perceptions of Campus Sustainability. *International Journal of Sustainability in Higher Education* 12 (1): 79–92.

Evans, J., R. Jones, A. Karvonen, L. Millard, and J. Wendler. 2015. Living Labs and Co-Production: University Campuses as Platforms for Sustainability Science. *Current Opinion in Environmental Sustainability* 16:1–6. https://doi.org/10.1016/j.cosust.2015.06.005.

Food Recovery Network. 2023. Food Recovery Network: Kentucky. https://www.foodrecovery network.org/kentucky#UL.

Helferty, A., and A. Clarke. 2009. Student-Led Campus Climate Change Initiatives in Canada. *International Journal of Sustainability in Higher Education* 10 (3): 287–300.

Henry, L. 2017. Understanding Food Insecurity Among College Students: Experience, Motivation, and Local Solutions Food Insecurity in College. *Annals of Anthropological Practice* 41 (1): 6–19. https://doi.org/10.1111/napa.12108.

Isenhour, Cindy, and Joshua Reno. 2019. Of Materiality and Meaning: Ethnographic Engagements with Reuse, Repair & Care. *Worldwide Waste: Journal of Interdisciplinary Studies* 2 (1): 1–8. https://doi.org/10.5334/wwwj.27.

Kuh, George D. 2008. *High-Impact Educational Practices: What They Are, Who Has Access to Them, and Why They Matter.* Washington, DC: Association of American Colleges and Universities.

Lassiter, L. E. 2005. *The Chicago Guide to Collaborative Ethnography.* Chicago: University of Chicago Press.

Martínez, F. 2017. Waste is not the end. For an anthropology of care maintenance and repair. *Social Anthropology* 25(3): 346–350. 10.1111/soca.2017.25.issue-3 10.1111/1469-8676.12436

M'Gonigle, M., and J. C. Starke. 2006. Minding Place: Towards a (Rational) Political Ecology of the Sustainable University. *Environment and Planning D: Society and Space* 24 (3): 325–348.

Mahon, Kathleen, Hannu L. T. Heikkinen, Rauno Huttunen, Tess Boyle, and Ela Sjølie. 2020. What is Educational Praxis? In *Pedagogy, Education, and Praxis in*

Critical Times, ed. Kathleen Mahon, Christine Edwards-Groves, Susanne Francisco, Mervi Kaukko, Stephen Kemmis, and Kirsten Petrie, 15–38. Singapore: Springer Nature.

Puig de la Bellasca, María. 2011. Matters of Care in Technoscience: Assembling Neglected Things. *Social Studies of Science* 41 (1): 85–106.

Puig de la Bellasca, María. 2012. "Nothing Comes Without its World": Thinking with Care. *The Sociological Review* 60 (2): 197–216.

UofL Sustainability. 2023. Composting. https://louisville.edu/sustainability/operations/composting.

Vélez-Ibáñez, Carlos G., and James B. Greenberg. 2000. Formation and Transformation of Funds of Knowledge Among U.S.-Mexican Households. In *Schooling the Symbolic Animal: Social and Cultural Dimensions of Education*, ed. Bradley A. Levinson. Rowman and Littlefield.

Waliczek, T., M. Holmes, and A. McFarland. 2016. The Relationship Between a Campus Composting Program and Environmental Attitudes, Environmental Locus of Control, Compost Knowledge, and Compost Attitudes of College Students. *HortTechnology* 26 (5): 592–598. https://doi.org/10.21273/HORTTECH03320-16.

CHAPTER 6

Fixing Coffeemakers and Giving Away Couches: Urban Economies of Reuse, Repair, Sharing, and Care in Germany

Petra Kuppinger

It is 9:50 am on a Saturday morning in May 2022. The first visitors are lining up in the lobby of a community center in Stuttgart, Germany, waiting for the monthly Repair Café to open. Some carry bags or boxes that contain coffee makers, blenders, printers, or vacuum cleaners, others hold a lamp or toaster in their hand; all broken items to be fixed by the expert volunteers who are standing in a circle on the other side of the glass wall that separates the lobby from a meeting room. The experts are discussing the last details of today's event. Soon, visitors sign in at a front desk and are then guided to an expert who will work with them to repair their object. Some visitors come to the Repair Café because they have a favorite item, old gadget, inherited appliance, or beloved toy. Most come because they are tired of hyper-consumerism, planned obsolescence, and the just-toss-it mentality. They wish to prolong the life of ordinary items and even

P. Kuppinger (✉)
Department of Sociology and Anthropology, Monmouth College,
Monmouth, IL, USA
e-mail: petra@monm.edu

P. Kuppinger (ed.), *Second-Hand Cultures and Economies of Reuse, Repair, Sharing, and Care*,
https://doi.org/10.1007/978-3-031-99875-1_6

minimally challenge hyper-consumerism. The experts, most are engineers with various specializations, volunteer to spend their Saturday sharing skills and helping people. Coffee, "saved" sandwiches, and baked goods that are provided by a food-sharing initiative are served. Sometimes there are homemade cakes on this small buffet.

The Repair Café is one of many projects and activities in Stuttgart where people challenge dominant consumer cultures and experiment with different types of consumption, material cultures, and socialities. Repair Café volunteers seek to contribute to a more livable and fair city. Visitors want to prolong their possessions' lifespans and, at least for now, save them from the incinerator (few landfills in Germany). Visitors and volunteers are part of a growing economy of reuse, repair, sharing, and care who confront dominant linear and unsustainable economies and work to create different practices, circuits, and communities. The cooperation of the Repair Café with food-sharing activists illustrates networks of alternative projects. The repair activists fix consumer items, participate in community making, public debates, and try to make the city a warmer and more cooperative place. Food sharing activists combat excessive food waste and save food, often from bakeries, that would be thrown away at the end of the day. They make this food available at specific pick-up locations in the city or give it to initiatives like the Repair Café. The question emerges whether such small projects, their networks, and cooperation can add up to larger social, ecological, cultural, and political change, or do they serve mostly as feel-good moments and spaces for their predominantly middle-class constituencies. Is the sphere of volunteer repair experts, food sharers, and other activists a small and relatively closed circle of dedicated individuals, or do they constitute a seed of more robust social and cultural change? How comprehensive and established does an urban field of reuse, repair, sharing, and care have to be to become an effective agent of change?

Recent years witnessed the proliferation of second-hand economies, networks of sharing and exchange, and creative efforts to extend the lifetime of products. There are growing landscapes of thrift stores, flea markets, yard sales, swapping events, neighborhood or Internet sharing networks/platforms, share boxes or shelves on sidewalks, foodbanks, food-sharing places, or repair cafés just to mention the most common ones. Based on ethnographic fieldwork in Stuttgart (since 2021), I introduce three exemplary elements of this growing landscape of reuse, repair, sharing, and care. I examine these projects' roles in larger social, ecological, economic, and political transformations. I examine the repair café, a charitable thrift store, and a city-run gifting website and analyze their work, motivations of organizers, workers, activists and/or customers, and their contributions to the creation of an ecologically and socially more sustainable and just city. Theoretically, I address questions of the role and potential of small projects of reuse, repair, sharing, and care in the remaking of consumer cultures and economies. I explore the synergies between projects and analyze their combined effects and potential. Can small projects and the networks in which they exist create more comprehensive and effective regenerative economies, and contribute to the transformation or transition of current unsustainable consumer cultures? How robust are the emerging networks of such projects? As initiatives, activities, and projects connect, can they produce tangible cultural, social, economic, and political transformations that are necessary to build regenerative and just future cities?

Economies of Reuse, Repair, Sharing, and Care

The study of second-hand cultures and economies of reuse, repair, sharing, and care has gained momentum in the social sciences (e.g. Appelgren and Bohlin 2015; Bohlin 2018; Ta 2017; Murphy 2017; Larsen 2019; Steward 2020). Nicky Gregson and Louise Crewe pioneered the study of complex second-hand markets in the UK, like car boot sales (1998). In *Second-Hand Cultures* (2003), a now classic text in second-hand studies, Gregson and Crewe insist on the importance of second-hand goods and markets and document the diversity and growth of this social and economic realm. Surveying the British landscape of charity and retro/vintage shops, they insist "second-hand worlds matter" (ibid., 1). They show how value is defined and created in second-hand markets and argue that second-hand exchanges create spaces and encounters that are "radically different from first hand spaces" (ibid., 4). They demonstrate the multilayered nature of second-hand exchanges that can include the "potential for critique, resistance even" and reflect movements of ethical, green, or sustainable consumption (ibid., 10).

Gretchen Herrmann (see this volume) is a veteran of research about garage/yard sales who produced groundbreaking work in the United States (e.g., 1997, 2004, 2015, 2019; Herrmann and Soiffer 1984). In the 1980s, before it was cool, Herrmann studied the material impact and the social, environmental, and community-building roles of garage sales in upstate New York. She examined garage sales as core elements in expanding second-hand cultures and economies. She dates the emergence of garage sales in the United States to the 1960s resulting from explosive production and consumption of goods when people started to discard still good and usable possessions to make room for new stuff (2019). Selling things cheaply to neighbors was a good solution, preferable to sending items to the landfill. By the 1980s, garage sales represented a "cultural style, a quasi-ideological rejection of consumerism" (Herrmann and Soiffer 1984, 411) for people who valued "recycling of goods" (ibid., 413) and its environmental impact (ibid., 415), when few thought in such terms. Herrmann describes garage sales as sites of meeting and community building (2004) and underlines the multilayered functions of sales that unfold in liminal spaces between private and public. These events allow people to unload excess belongings to declutter their homes or buy more stuff. They generate small profits for sellers and allow shoppers to acquire goods cheaply. Herrmann highlights the social aspects of garage sales where neighbors can get to know each other without having to open their homes (2015). Other observers of activities in economies of reuse, repair, sharing, and care similarly emphasize their community-building elements (e.g., Chiveralls 2018; Albinsson and Perera 2012).

Margaret Crawford adds that before the time of garage sales, extra possessions were "passed on through kinship and friendship networks" as "hand-me-downs" (ibid., 25) or given to charitable organizations for their occasional rummage sales (ibid., 26). With growing piles of stuff and hoping to make a small profit off excess possessions, people started selling them out of their homes/garages, while having fun with their neighbors. With growing volumes of household goods in the 1970s and the quest to purge things to purchase more, Crawford notes that by the 1980s garage sales had become part of "mainstream American life" (ibid.).

Robert Crocker and Keri Chiveralls (2018) examine reuse and related practices that counter hyper-consumerism. They note that in an "accelerated world," slow practices and those that counter hyper-consumerism and environmental degradation are paramount, because they challenge dominant capitalist regimes and seed solutions, propose change, and outline circular and regenerative economies (Crocker & Keri 2018a). Reuse is more than "the recovery and circulation of second hand goods" (ibid., 2), it includes the "reuse of what was once regarded as unusable and unsalable, including the 'upcycling' of unmarketable wasted products such as food, broken metals, timber, e-waste and even toxic chemicals" (ibid.). Reuse challenges dominant definitions of waste and consumption.

The broader realm of reuse requires knowledge and careful engagement with all materials. While "ordinary" second-hand circuits of clothes, furniture, appliances, and toys remain central, it is crucial to question notions of garbage, develop practices to retrieve materials, and rewrite definitions of waste.

Being aware that reuse often reflects contexts of financial necessity and remains of vital importance for marginalized groups, Crocker and Chiveralls and their contributors insist on the political nature of reuse, repair, or recycling projects. Dedicated practitioners see themselves at the forefront of political transformations and transition movements, representing material aspects of political ideas and philosophies that combat environmental degradation and social injustice (Fraser 2018). Seemingly insignificant "niche" practices or innovations have the potential to take hold and are "window[s] of opportunity" that can generate larger transformations (Waller et al. 2018, 191). When critical activities are initiated, it is impossible to predict which ones will inspire others and engender change. Seeds of change flourish in unexpected circumstances (Kuppinger 2021a, b; Fessenden 2021). "Niche" or seeding practices are vital in the struggle for ecological and social sustainability and justice. Contributors to Crocker and Chiverall's volume highlight activities such as transforming urban wastelands, composting, textile upcycling, and reusing building materials where individuals or groups engage in practices that are drawing in others or might do so in the future (e.g., Lee 2018; Waller et al. 2018; Crocker 2018; Chiveralls 2018). They work on material improvements while pursuing larger political goals.

Cindy Isenhour and her collaborators, especially Brieanne Berry (see this volume) study second-hand stores, work, circuits, and exchange networks in rural Maine and their connections to larger markets. Isenhour and Berry analyze individual efforts and small projects in long-established reuse economies in rural Maine (flea markets, thrift stores, yard sales, swapping; Berry et al. 2019; Isenhour and Berry 2020). These activities do not put sustainability on their banners, and practitioners do what they always did (based on ideas about frugality and thrift), or act because they have few other economic choices. Isenhour and Berry do not simply applaud the ecological impact of people's activities to salvage, repair, redistribute, or sell second-hand items, but analyze inequalities and dynamics of marginalization and exclusion in the rural second-hand economy. They identify rural second-hand circuits that provide income or bare survival for economically marginalized individuals who labor to find, clean, repair, and present items. This invisible and underpaid labor feeds into profitable networks of re-/selling and provides "treasures" to wealthy dealers and urbanites who do not labor to locate or restore items. Rural workers at the bottom of this second-hand labor chain produce value by finding and restoring pieces in what others see as "depleted" rural landscapes. They create value from trash, discards, and forgotten items. Isenhour and her collaborators illustrate how notions of thrift, frugality, and care are interwoven with those of antiques, commercialization, and rural marginalization. They stress the environmental impact of (local) circular economies but question the "increasing commodification of discards" and the extraction of valuable goods from rural circuits, their profitable use by outside agents, and insertion into upscale vintage or antique markets (2019). They suggest linking rural and urban circular economies in a just manner, a process underway on second-hand and/or sharing platforms like ThredUp or Facebook Marketplace. Isenhour and Berry argue that (marginalized) local economies remain fundamental to larger circular economies, which, because of their disadvantaged economic position, shoulder disproportionate risks. Often written off as "places that do not matter" (2019, 112), rural economies of reuse, repair, and thrift offer valuable lessons about saving money and reducing waste and illustrate the creation and maintenance of community and meaning in small towns.

Cindy Isenhour and Joshua Reno (2019) explore emerging cultures and economies of reuse, repair, and care as a complex cultural field at the intersection of established frugal cultures and more recent anti-consumerism. They identify care as an often-forgotten aspect of the material world. Care indicates a different, more reflective, and dedicated relationship to material things and the world at large. Reuse and repair imply a thoughtful handling of things that transcends the buy-use-discard linearity of contemporary hyper-consumerism and connects to older economies and practices. It engenders new practices that imply circularity, resistance, and make statements against capitalism.

Adam Minter (2019) takes a trip around the globe highlighting second-hand activities, practices, and sites. In Minneapolis, he reflects about the quantities of excess goods moved out of site to storage units. He looks at Japanese decluttering practices and the business of cleaning the houses of the deceased. He ponders about the "flood" of goods that Goodwill and similar stores in the United States are struggling with every day. He notes that Goodwill's more than 3000 stores in the United States and Canada divert 3 billion pounds of materials annually from landfills. Yet this accounts for a mere 3 percent of the "clothes, furniture, and miscellaneous durables tossed out" (2019, 48–49). He visits antique malls in the United States and traces Japanese discards that are resold in the Philippines, Pakistan, Nigeria, and elsewhere where buyers appreciate Japanese products. Hamaya, a company specialized in the global resale of used Japanese goods, reached sales of $1 billion/year in the 2010s (ibid., 93). Hamaya is only one of many similar companies that export used goods from Japan (ibid., 95). Minter describes the individualized second-hand trade between Arizona and Mexico where small traders ferry goods in trucks across the border. Minter notes that this cross-border trade in American discards is "one of North America's most environmentally sustainable businesses, but nobody notices it unless they're in it" (ibid., 108). In Ghana, Minter visits workshops where experts fix TVs and other discarded appliances and electronics from Europe, the United States, and East Asia. These repair experts can bring any appliance back to work if parts are available. Parts are often sourced via companies in the United States whose sole business is to break down TVs and sell parts across the world (ibid., 230). Minter provides a glimpse at global second-hand circuits. He points to vast inequalities that mark this profitable used goods market, such as who does the dirty work of sorting second-hand goods in the United States or works in the small-scale cross-border trade. He points to the destructive reality of dumping mountains of second-hand clothes in Africa (see Brooks 2019).

In 1987, long before sustainability studies emerged, Douglas Harper wrote a brilliant, profound, and poetic study of the work and life of Willie, a repairman and master of many trades, who ran an auto and farm equipment repair shop in upstate New York. Practicing preindustrial modes of working and living, Willie repaired vehicles and machinery to prolong their lifespan, add features, and improve elements beyond their planned functions. Harper describes Willie as a genius mechanic and the center of a vibrant social network based on complicated and long-term dynamics of work, community, and exchange. Willie did not make riches but remained "in control of his time and energy" (Harper 1987, 4). He commanded impressive skills and owned a massive supply of "stuff" (anything from old cars to odd pieces of wood or metal) that might come in handy one day. In Willie's world, little counted as trash. Harper notes "Willie's lived definition of use value" stood in stark contrast to "general cultural values of planned obsolescence" (ibid., 9). Harper explains, "Willie defines himself through his work and how his work leads to a sense of acting and being in the community" (ibid., 11). Fixing things, especially old farm equipment,

Willie repaired a disappearing world of small family farms and stemmed the tide of threatening rural disenfranchisement and poverty. None of this paid much but provided Willie with a sense of self, being in the world, and communal cohesion. Willie turned "the leftovers of the world into replacement parts and even new machines" (ibid., 12).

Willie was a brilliant specialist who redesigned machinery, and a man of vast and embodied knowledge guided by a vision of the world were materials, technical expertise, and communal well-being were inextricably linked. Repair was not a few steps outlined in a manual, but a complicated rethinking of interwoven technical and social functions. Willie's work was the product of his experience and that of those before him. He sought to bring defunct items back to function and rebalance the social universe. Willie was "outdated" in the world of a rapidly growing toss-it mentality in the 1980s. He cared little for this world but focused on social, material, and environmental well-being. Repair was part of a holistic lifestyle that strove to maintain harmony in the world. Willie did not engage in (compartmentalized) sustainability and community-creating projects; he dedicated his life to these issues without needing mention.

In recent years, more researchers have explored the world of repair (Strebel et al. 2019; Martínez and Laviolette 2019; Graziano and Trogal 2019; Schmid 2019). Lara Houston studied the work of cell phone repair in Kampala, Uganda. Most of the cell phones there used imports. Houston entered a globalized "repair world" (2019, 133) where experts in Kampala maneuver complex worlds of technology and exclusion (power of authorized dealers vs. independent technicians). Denied access to manuals and workshops, these technicians rely on cyber networks of similar experts across the globe who post experiences, advice, and solutions and share software. Houston highlights the impressive knowledge, diagnostic tools and technology, and cooperation of highly skilled workers in Kampala's tiny workshops who maneuver their marginalized position in the global tech world. Repair work in Ghana (Minter 2019) and Uganda (Houston 2019) highlights ecological benefits but also illustrates glaring global inequalities. Repair is an important and worthwhile sustainable activity but remains marked by inequality. We are far from a world like Willie's where one repair expert could repair material items and foster communal justice. Repair has long moved from a holistic exercise to a materially limited (but still worthwhile) task. Repair and access to repair knowledge reflect global inequalities and exclusion.

In the early twenty-first century, economies of reuse, repair, sharing, and care unfold in landscapes where necessity, lifestyle choices, and political activism intersect. Individuals have appliances repaired because they cannot afford new ones or because they want to confront hyper-consumerism. They buy second-hand clothes because they must or because they chose to challenge consumption, especially fast fashion. Looking at second-hand, repair, and related projects, the question emerges whether they remain isolated or develop connections and synergies that engender robust economies of reuse, repair, sharing, and care. Are there new lifestyles that are at least vaguely reminiscent of Willie's holistic world of repair, reuse, communal cohesion, and justice? Are thrifters more likely to ride their bicycles to work? Do those who fight the planned obsolescence of their printer avoid the use of plastic or frequent food-sharing places? Are visitors to the Repair Café more likely to buy thrifted clothes or buy unpackaged groceries? Do individuals engaged in multiple activities seed political change?

Repair Café

The Repair Café opens once a month in a community center in a middle-class inner-city neighborhood known for its green political leaning. Sometimes, upon invitation, the Repair Café "opens shop" in other community centers or public locations. Opening hours are 10 am to 3 pm, and about 30–40 visitors come on an average Saturday. Most repairs are appliances and electronics. There is the occasional bicycle or even lawn mower, and depending on availability of experts, textile repairs are offered. During my fieldwork in 2021/22, I did the textile repairs. Since this station had been vacant for some time, there were only a few (outside) visitors to the station, but the volunteers kept me occupied. I fixed jeans, shirts, t-shirts, sweaters, pajamas, and bed sheets for them and some visitors. The experts are dedicated to sustainable practices and like the idea of fixing and mending in their own lives.

The Repair Café idea originated in Holland in 2009 and has since spread across many countries.[1] The idea is to combat waste and planned obsolescence. The movement addresses hyper-consumerism, works toward a more sustainable world, seeks to bring people together, and build communities. Within a few years, thousands of Repair Cafés emerged in Europe, the United States, and Japan. In Germany, hundreds of Repair Cafés operate in community centers, churches, and libraries. In the

Stuttgart region, there are tens of such projects. The website of the international umbrella organization of Repair Cafés states there are currently 3005 repair cafés globally, where over 45,000 volunteers fix about 54,000 items every month.[2] The international and national umbrella organizations help groups with new projects and assist with legal questions like the nonliability agreement that visitors sign. The philosophy of Repair Cafés is to help people repair their items and teach them details for future repairs. Visitors sit with the expert volunteers and learn some repair practices. Items cannot be dropped off for repair.

The overall repair success rate of the Stuttgart Repair Café is between 50 and 80 percent. Some appliances are easily fixed by replacing small parts or reconnecting a cable. Volunteers frequently send visitors to nearby hardware stores to pick up parts and together they insert them into appliances. The atmosphere at the Repair Café is friendly and relaxed. Some visitors bring their children to teach them about fixing things and understanding the value of repairing items. Some repairs are tricky and take hours. The point of the Repair Café is not just to quickly repair things, but to work together in a communal and cooperative atmosphere. Visitors and helpers socialize, and interesting discussions unfold, often focusing on consumerism and sustainability. Some visitors come to accompany the owner of a broken item and wander between tables, watch repairs, and participate in discussions. Many visitors watched and commented on textile repairs or engaged in discussion about anything from fast fashion, veganism, or waste reduction to their inquiries about sustainable practices and projects in the United States.

On my first day at the Repair Café in October 2021 (I did not have much to fix), a woman, Anna, came with a decorative table lamp, which she had bought second-hand that did not turn on. The volunteer, Michael, took the switch apart and said that somebody had inserted an unsuitable dimmer. He sent Anna to the hardware store to buy the correct part. When Anna returned, the part was too large for the switch. Michael apologized for the misunderstanding and sent Anna to exchange it. Anna returned with a perfect piece. Over the course of almost two hours, Michael and Anna brought the lamp back to work, I had a very pleasant conversation and coffee and cake with Anna. While the lamp was old and not "worth" that much, Anna loved it and hoped it would serve her for some years. Michael was happy to help Anna and teach her details about the workings of light switches, how they can be messed up, and fixed again. This encounter was not only about fixing a lamp. If Anna only wanted a light for her living room, it would have been easier to buy a cheap lamp at a big box store. Neither Anna nor Michael considered this a solution in a world of overconsumption and excess waste. They took the time on a beautiful Saturday afternoon (the encounter took place in the beautiful garden space of the center which the Repair Café uses in good weather) to fix this lamp.

On another beautiful Saturday in the spring 2022, a couple in their 70s came with a 1970s food processor. Arriving at opening, they were sent to Ulrich, a retired engineer—also in his 70s, who was waiting for "work" in the garden. I sat at the next table with the sewing machine waiting for ripped textiles and observed their interaction. The husband, Karl, explained that they have had this machine forever and wanted to continue using it, because it was good and for sentimental reasons. He explained that he had partially disassembled the machine but got stuck at a certain point (he had an engineering background, he added). Ulrich and Karl took the food processor apart, neatly placing the pieces next to the processor on the table. It was a matter of professional honor for the two retirees to get this processor back in shape. Gerlinde, Karl's wife, sat and watched them work and discuss the processor. Occasionally she left to get food or watch other repairs. Barely aware of their surroundings, Ulrich and Karl worked on the food processor for more than two hours and their joined efforts brought the appliance back to life. Karl and Gerlinde were extremely pleased to see their treasure working again. On the way out, like most visitors, they left a donation in the box (for tools, materials, and coffee). Karl and Gerlinde looked like they could afford a new food processor, but their hearts were attached to the old one and they insisted on not buying new stuff, when things could be fixed. At least one should try.

Not everybody is as lucky as Anna, Karl, and Gerlinde. One visitor brought a 1960s film projector. He had inherited boxes of old family reels, and this was the old projector with which to watch the films, otherwise the heirlooms were useless. Two volunteers joined the visitor set on reopening the view onto decade-old memories. They took the projector apart, checked cables and connections but could not identify the mistake. Together with other experts whom they called to the table, they spent more than two and half hours trying to fix the projector. Unfortunately, they could not help, and the man went home with his still dysfunctional projector.

Thrift Store

Fairkauf is a large charitable second-hand store located on the border between the residential and industrial parts of an inner-city quarter in Stuttgart.[3] The store (run by a Catholic charity) occupies a defunct industrial facility that includes a yard and parking lot. The premises face an urban highway, an apartment building, a school, and industrial and commercial facilities across the highway. The main building includes a ramp for receiving donations, spaces for sorting donations, storage spaces, and a large two-story store. There are more storage facilities in the backyard. Fairkauf takes donations of clothes, household items, toys, appliances, and furniture to be sold at low prices to those in need or those who seek to avoid consumerism. Like similar institutions in other countries, Fairkauf provides jobs for long-term unemployed individuals or those marginalized in the first labor market who work there in cooperation with the federal employment agency. There are volunteers (fewer than before COVID) and some individuals who perform mandated community service. A professional full-time staff of administrators and experts (textile, woodwork) oversees the store. Fairkauf used to offer pick-up services for furniture or household appliances but does not currently provide these because of labor shortage.

People drop off donations at the ramp by the parking lot where workers sort and channel goods to the appropriate stations in the building. Upon sight or asking donors, workers put bags and boxes onto carts designated for children's or adult clothes, home textiles, toys, appliances, and other categories. Full carts are delivered to the sorting stations, where workers check every item whether it is good for sale, can be recycled, or should be discarded. There are stations for household items, appliances, children's clothes, adult clothes, or home textiles.

Fairkauf is always busy. About 20–30 workers receive, sort, clean, and sell endless streams of clothes, shoes, books, appliances, or furniture. A similar number of customers is usually in the store searching, trying on clothes, and looking for other treasures. Clothes are the single most donated goods. Depending on the amount of donations and available workers at the textile sorting stations, ramp workers carry sacks and boxes right to the sorting tables. I spent hours at these tables, often alone, when workers were in short supply, or sometimes with another worker. Most textile sorting happens here or at smaller (often part-time) stations for shoes, belts, bags, accessories, or seasonal items. Adult clothing is sorted by season and type (shirts, pants, jackets, etc.). Women's clothing outweighs men's by about 9:1. In the basement, there is a sorting station for household and decorative items. One part-time worker exclusively works with Christmas and other seasonal items throughout the year. Even tiny items are readied for sale. For example, Christmas knick-knacks are put in small bags with similar items and priced at 50 cents. These bags are stored in the basement until their next season.

On a "good" day, especially in the late afternoon, donations arrive minutes apart from each other. Toys, appliances, dishes, home decoration, and seasonally appropriate clothes are sorted and immediately brought to the store where workers put them in their assigned places. Clothes can make it from the ramp to the store in only a few hours. About 60 percent of the two sales floors is reserved for clothes and there are sections for furniture, toys, dishes, home textiles, and books. Walking through the well-stocked sales floor, one can witness the excesses of consumer society. To fill 2000 square meters at all times with plentiful donations bespeaks hyperconsumption. Some clothes (and other items) arrive unused with price tags still attached. In a wealthy city like Stuttgart (world headquarters of Mercedes-Benz, Porsche, and Bosch), donated clothes also include some high-end and luxury pieces. In one afternoon of hanging presorted clothes, I encountered items by local and global upscale brands like Armani, Escada, Hugo Boss, Marc McCain, or Luisa Cerano.

Fairkauf tries to produce as little garbage as possible partly because the city charges them for every industrial-size container that is picked up. Stained, ripped, or otherwise unusable clothes are sold to dealers who resell or shred them. While sustainability efforts are not the core purpose of Fairkauf, there are small signs on the sales floor that explain that second-hand clothes equal sustainable practices. Signs state how much water is used in the production of a T-shirt and how customers contribute to a better world by shopping second-hand. Every day tons of goods enter the backdoor and exit through the front door of Fairkauf. Huge quantities of materials are diverted from incinerators and their lifespan is extended. The ecological contribution of Fairkauf is enormous, but Fairkauf does not make loud claims to sustainability.

Fairkauf attracts diverse customers. Many come because of economic necessity. Others consider second-hand cool. Some buy there because they do not want to contribute to overconsumption and prefer to buy some or most of their necessities second-hand. Customers represent a broad variety of demographics, like refugees from Syria and Ukraine (since 2022).

They hold special cards from the city, which entitle them to a 20 percent discount. There are students from a nearby trade school and middle-class and middle-aged to older ethnic Germans (mostly women) searching for bargains and crafting materials. It is hard to estimate who is who but in a survey I conducted there in the summer of 2022, many customers indicated that ecological motivations played an important role for them. Regardless of individual necessities and motivations, Fairkauf provides a huge service to the city and planet. It offers those who do not want to buy new items an impressive selection of clothes, shoes, dishes, and appliances. Whether bought because of financial necessity, for the coolness factor, or to avoid wasteful consumption, Fairkauf customers save hundreds of tons of materials from the incinerator. They counter hyper-consumption and strengthen economies of reuse, repair, sharing, and care.

Gifting Website

Stuttgart's municipal Department of Waste Management (*Abfallwirtschaft*) runs a website called Gifting Market Stuttgart (*Verschenkmarkt Stuttgart*) where residents can post things to be gifted.[4] Objects on this site must be free; nothing can be sold. The city offers this online service to residents at no cost (like many other German cities). The website aims to help people create space or declutter their homes, limit waste, save resources, and find needed free items. It is not surprising that the Department of Waste Management runs this site as it benefits from having to pick up and process less (often bulk) waste. The service of connecting citizens with excess stuff with those who need things is a welcome bonus. The website is well known. If a person wants to sell their goods, they use sites like *Ebay Kleinanzeigen* (local sales on eBay).

The Gifting Market is widely used, and about 15–20 new items are posted daily. Individuals can list anything they want to gift from a book, a jacket, or three plates, to couches or bedroom sets. On a random day (February 13, 2024), individuals offered a glass bowl, vases, an L-shaped couch set with coffee table, a mattress, a bicycle, printer, desk, bed, bags, small dining table with chairs, or another couch with coffee table. Those who need things can post their searches. On this day, people asked for an e-bike and binoculars. Those interested in posted items contact those who offer them. Gifting is conducted on a first come, first served basis. For popular items, one has to move fast which means arranging for a pickup time as fast as possible; usually no later than the next weekend. If unclaimed, items can stay on the website for a month.

Attractive objects disappear within hours, some stay for days or the full month. Individuals who want, for example, a couch check daily until the right model appears. Pick up procedures are (for the safety of everybody involved, like on Facebook Marketplace or *Ebay Kleinanzeigen*) for items to be brought to the front door/apartment door by the giver. The recipient checks that the object is in order and as depicted in the picture. If they agree to take it, they should take it right away or arrange for swift dismantling and transportation for larger items. Considerable quantities of dishes, home textiles, toys, and furniture change hands via this website and tons of materials are diverted from the waste stream every month.

With a friend who furnished a new apartment, I went in her neighborhood to pick up a large and heavy floor lamp. At the appointed time, we rang the doorbell of the giver and a woman came downstairs and told us to wait until she retrieved the lamp from basement storage. She plugged in the lamp in the hallway to show that it worked. She told us that her husband had never really liked the lamp much and that is why they were giving it away. She warned us that the lamp was slightly shaky, but she thought this could be fixed. Together, we carried the lamp to my friend's apartment (one carrying at the top, the other at the foot of the lamp). At home, with a screwdriver, we fixed the slightly wobbly stand and the lamp was like new.

Like in many second-hand transactions, stories about the gifted item might be exchanged at the doorstep (the husband not liking the lamp), like where is it from, who owned it before (heirlooms), or what special meaning it had to this or a previous owner. While these encounters are brief and anonymous transactions, many seek to hand over things with a story, hoping for them to find a good home for their story to continue. While items are unwanted in their current home, givers know that others will use them and might continue their stories. Knowing this, and that things will not go to the incinerator (for now) makes it easier to let go of them.

Does It Add Up?

The Repair Café, Fairkauf, and Gifting Market are part of an urban landscape of projects that address ecological, economic, and social issues. There are flea markets, share websites, and share shelves/boxes on streets where people leave their discards. There are swap parties and food sharing shelves and apps. There are commercial projects, like an upcycling fashion brand and upcycling artists (Kuppinger 2021c). There are unpackaged stores (ibid.) and a For-Free/Foodsharing Café (patrons decide whether and what to pay for food and drink),[5] just to mention some elements of a growing economy of reuse, repair, sharing, and care in Stuttgart (cf. Schmid 2020). The city has a produce market (*Wochenmarkt*) on the central market square, which has been in operation since 1304. Since then, regional farmers have offered their produce every Tuesday and Saturday, since 1775 also on Thursday.[6] For more than 700 years, this market has connected the city to the countryside and allowed citizens to buy regional produce. The market provides a consumer experience and relationships that represent the survival of elements of a more holistic world where people like Willie lived and practiced their trade in the context of communities where social and material elements were interwoven. The market provides produce and community, and shoppers (often for decades) buy produce from the same family farm. The absence of a seller or customer might spark concerns about their well-being. Many shoppers bring their own produce sacks or have their purchases emptied straight into their bags to avoid the use of plastic. Whether as continuous elements or instances of rethinking and revival, institutions like the *Wochenmarkt* illustrate a sustainable future that combines the past and future.

Looking at these places and practices, some questions emerge. Do they add up to larger transformations? Do a fixed coffeemaker, thrifted shirt, gifted couch, and regional apples in a reusable bag really make a difference? Materially, they make minute differences, as they do not (yet) travel to the incinerator or do not travel far (apples) which is a start and sets a symbol. And there are more repaired appliances or thrifted items in Stuttgart and beyond. Each project and activity contributes, in a small way, to a socially and ecologically more sustainable and just city. These material acts foster debates about their meaning, results, and political relevance. As a sign at Fairkauf indicates, a thrifted T-Shirt saves 2500 liters of water versus the production of a new shirt. This is water taken from people halfway across the globe. The thrifted T-shirt does not end global injustices but draws attention to and challenges them in a tiny gesture. It is a seed that inspires others or has its new owner consider additional sustainable practices. Each small act can be a stepping-stone for more practices and engagements. Participants in voluntary projects and workers in commercial projects are often busy and seem to work in isolation from each other. Yet their combined efforts outline larger possibilities and future economies. Their work merges in the practices of dedicated individuals who partake in projects and whose lifeworlds link the Repair Café, Fairkauf, the Gifting Market and others.

In June 2022, I conducted a survey among visitors of the Repair Café. Twenty-two individuals answered my questionnaire: 11 men and 11 women between the ages of 24 and 80. I asked them about their visit to the Repair Café and the objects they brought. I included questions about visitors' sustainable everyday practices and engagement in economies of reuse, repair, sharing, and care to understand how their visit to the Repair Café related to other activities and similar places they frequented (including Fairkauf and the Gifting website). Eleven individuals had shopped at Fairkauf and five had used the city's Gifting website. Eleven had been to Oxfam, four had shopped at *Zora* (female run second-hand store), or the *Second-hand Sozialkaufhaus Cannstatt* (6 people; second-hand "department store")—all popular charitable second-hand venues. Four individuals shopped at the Unpackaged grocery store, and four had visited the For-Free/Food-Sharing Café. Nine people frequented flea markets and eleven had bought things from *Ebay Kleinanzeigen* (local small sales). When asked about what other sustainable activities they engaged in, 17 individuals noted, "they avoid unnecessary consumption," 10 stated they buy "a lot of used items." Thirteen people said they avoided the use of "unnecessary plastic item." Seven stated that they did not own a car; 10 routinely rode a bicycle, and 13 donated things to places like Fairkauf. Six people checked that they crafted or upcycled things. Seven people said they were vegetarians and two participated in food-sharing activities.

This (small) sample illustrates the interwoven nature of Stuttgart's economy of reuse, repair, sharing, and care. People participate in different contexts, and sustainable practices do not occur in isolation. Individuals combine them. A person might ride their bicycle, be a vegetarian, repair their printer, and buy thrifted clothes. There is no one or perfect way to participate in economies of reuse, repair, sharing, and care. People pick practices and combine them in ways that are best for them and effective in their small ways and contexts. I was not surprised to run into a co-repairer at the unpackaged store or meet a friend who does much upcycling at the Repair Café. There are shared territories and shared practices that unite people who want to contribute to creating a more sustainable and just city and world. Will their efforts be enough to create a different urban economy? Not (yet) now, it seems, but it is worthwhile to support their growth and observe their future. If they continue to grow and spread, they will produce tangible transformations and political change. Seeds grow and steps add up. Since 2021, the Repair Café has had several requests for help and advice from groups that want to set up their own events. Raupe Immersatt, the food-sharing and for-free café opened in 2019 and just celebrated its fifth anniversary (Bosch 2024). More share-shelves and boxes pop up all over the city.

The material effects of the activities discussed in this chapter remain small. However, fixing coffeemakers, thrifting shirts, and giving away couches are crucial seeds for social and economic transformations. These niche practices and innovations are windows of opportunity for cultural change. We are far from the holistic social and material universe that Willie inhabited, but this should not keep us from reconsidering some of its elements and creatively putting them back to use. The point is not to return to the past but to take inspiration from meaningful ideas and practices and create a sustainable and just future. Small practices can foster cultural changes where more people challenge and reject hyper-consumerism and adjust their lifestyles and patterns of consumption accordingly.

The survey among Repair Café visitors showed that people who fix their possessions also thrift or frequent an unpackaged grocery store. Engaging in one practice and moving in related circles makes them consider other practices and solutions. They thrift, ride their bicycle, avoid plastic, or become vegetarians. While there is no preferred trajectory or perfect assortment of practices, they are linked to other similar ones. Adoption of one might lead to others. Small seeds (or whatever caught people's attention first) are important as they seed other changes when people join larger networks and frequent more similar places. One thrifted T-Shirt can be the start of more radical lifestyle choices and political involvement. People have to start somewhere.

Concluding Remarks

Like other cities, Stuttgart is home to a growing economy of reuse, repair, sharing, and care. People participate for varying reasons, and with different motivations in diverse contexts. Some are marginally involved in these contexts. Others are passionate and very knowledgeable about these economies, partake in many activities, and advocate for their expansion. They change their lifestyles, volunteer for change, and seek to engage others in debates about projects and activities. On the material level, the Repair

Café, Fairkauf, and the Gifting Market reduce waste. On the cultural level, they advertise different ways of consumption and alternative lifestyles (second-hand clothes as a value). On the political level, they advocate for larger change (e.g., the right to repair). They challenge hyper-consumerism and outline different ways of living, consuming, using, and sharing. They advocate fair and just consumption and highlight injustices. The Repair Café intends to connect people, teach skills, and provide a space for meeting and (political) debates. From a material perspective, the Repair Café contributes the least. With a success rate of 50–80 percent per session of about 40 visitors, they are able to save 20–30 items/month. These numbers need to be seen in a global context where Repair Cafés make a robust contribution. While diverting goods from the incinerator is a central goal of the initiative, it is equally important for volunteers that they offer comfortable spaces for exchange and debate among volunteers and visitors. Monthly events attract similarly minded individuals, encourage the exchange of ideas, repair knowledge, and information about other projects and resources in Stuttgart and beyond. Fairkauf turns over huge quantities of materials and makes them available to those with lesser means. While this is not a long-term solution for overconsumption and economic exclusion, it works for the time being, until we devise more sustainable and fair economies. Fairkauf does not connect donors and customers and there is usually no exchange of an item's story, as often happened when things change hands via the Gifting Market. Like Fairkauf, the Gifting Market diverts sizable quantities of materials from the waste stream. It connects those who want to give and those who want to take. The Market provides fleeting encounters between previous and future owners of things.

The Repair Café, Fairkauf, and Gifting Market are elements of a growing urban economy of reuse, repair, sharing, and care. None of them existed 30 years ago. They result from, symbolize, and simultaneously challenge capitalist hyper-consumption. They outline sustainable practices and futures and provide spaces for urbanites to experiment with and develop more sustainable lifestyles and fight for equitable cities. Their current impact is limited, but seeds are growing and attracting more people.

Notes

1. www.repaircafe.org
2. www.repaircafe.org; numbers for February 2024.

3. Fairkauf is the real name of the store. I use this name with the consent of the store's administration. I have written elsewhere in more detail about this thrift store (Kuppinger 2023).
4. www.verschenkmarkt-stuttgart.de
5. https://www.raupeimmersatt.de/
6. https://www.stuttgarter-wochenmaerkte.de/wochenmaerkte/geschichte/

References

Albinsson, Pia, and Yasanthi Perera. 2012. Alternative Marketplaces in the 21st Century: Building Community Through Sharing Events. *Journal of Consumer Behavior* 11:303–315.

Appelgren, Staffan, and Anna Bohlin. 2015. Growing in Motion: The Circulation of Used Things on Second-Hand Markets. *Culture Unbound* 7 (1): 143–168.

Berry, Brieanne, and Cindy Isenhour. 2019. Linking Rural and Urban Circular Economies Through Reuse and Repair. *Journal for the Anthropology of North America* 22 (2): 112–114.

Berry, Brieanne, Jennifer Bonnet, and Cindy Isenhour. 2019. Rummaging Through the Attic of New England. *Worldwide Waste: Journal of Interdisciplinary Studies* 2 (1): 6.

Bohlin, Anna. 2018. "It Will Keep Circulating": Loving and Letting Go of Things in Swedish Second-Hand Markets. *Worldwide Waste: Journal of Interdisciplinary Studies* 2 (1): 3.

Bosch, Julia. 2024. In diesem Stuttgarter Café entscheiden die Gäste selbst über Preise. *Stuttgarter Zeitung*, 7 June.

Brooks, A. 2019. *Clothing Poverty: The Hidden World of Fast Fashion and Second-Hand Clothes.* London: Bloomsbury Publishing.

Chiveralls, Keri. 2018. Reuse in Earthship Construction: Reclaiming the Past to Shape the Future. In *Subverting Consumerism*, ed. Robert Crocker and Keri Chiveralls, 204–237. London: Routledge.

Crewe, Louise, and Gregson Nicky. 1998. Tales of the Unexpected: Exploring Car Boot Sales as Marginal Spaces of Contemporary Consumption. *Transactions of the Institute of British Geographers* 23 (1): 39–53.

Crocker, Robert. 2018. Acceleration, Consumerism and Reuse: A Changing Paradigm. In *Subverting Consumerism*, ed. R. Crocker and K. Chiveralls, 11–34. London: Routledge.

Crocker, Robert, and Keri Chiveralls, eds. 2018. *Subverting Consumerism: Reuse in an Accelerated World.* London: Routledge.

Crocker, Robert and Keri Chiveralls. 2018a. Introduction. In *Subverting Consumerism*, ed. R. Crocker and K. Chiveralls., 1–10. London: Routledge

Fessenden, Sarah. 2021. Food is a Right Not a Privilege; Punk Anarchism, Ephemerality, and Seeding Change in Vancouver. In *Emergent Spaces*, ed. Petra Kuppinger, 227–246. New York: Palgrave Macmillan.

Fraser, Kim. 2018. ReDress: Maximizing Component Reuse for Fashion. In *Subverting Consumerism*, ed. Robert Crocker and Keri Chiveralls, 166–185. London: Routledge.

Graziano, Valeria, and Kim Trogal. 2019. Repair Matters.*Ephemera: Theory and Politics in Organization* 19(2):203–227.

Gregson, Nicky, and Louise Crewe. 2003. *Second-Hand Cultures*. Oxford: Berg.

Harper, Douglas. 1987. Working Knowledge: *Skill and Community in a Small Shop*. Chicago: University of Chicago Press.

Herrmann, Gretchen M. 1997. Gift or Commodity: What Changes Hands in the US Garage Sale? *American Ethnologist* 24 (4): 910–930.

Herrmann, Gretchen M. 2004. Haggling Spoken Here: Gender, Class, and Style in US Garage Sale Bargaining. *The Journal of Popular Culture* 38 (1): 55–81.

Herrmann, Gretchen M. 2015. Valuing Affect: The Centrality of Emotion, Memory, and Identity in Garage Sale Exchange. *Anthropology of Consciousness* 26 (2): 170–181.

Herrmann, Gretchen M. 2019. Reminiscence and Recompense: Reuse and the Garage Sale. *Worldwide Waste: Journal of Interdisciplinary Studies* 2 (1): 2.

Herrmann, Gretchen M., and Stephen M. Soiffer. 1984. For Fun and Profit: An Analysis of the American Garage Sale. *Urban Life* 12 (4): 397–421.

Houston, Lara. 2019. Mobile Phone Repair Knowledge in Downtown Kampala: Local and Trans-Local Circulations. In *Repair Works Ethnographies*, ed. Ignaz Strebel, Alain Bovet, and Philippe Sormani, 129–160. Singapore: Palgrave Macmillan.

Isenhour, Cindy, and Brieanne Berry. 2020. "Still the Good Life": On the Value of Reuse and Distributive Labor in 'Depleted' Rural Maine. *Economic Anthropology* 7:293–308.

Isenhour, Cindy, and Joshua Reno. 2019. On Materiality and Meaning: Ethnographic Engagements with Reuse, Repair & Care. *Worldwide Waste: Journal of Interdisciplinary Studies* 2 (1): 1.

Kuppinger, Petra, ed. 2021a. *Emergent Spaces. Change and Innovation in Small Urban Spaces*. New York: Palgrave Macmillan.

Kuppinger, Petra. 2021b. Introduction: Emergent Spaces: Seeding Change and Innovation in Small Urban Spaces. In *Emergent Spaces*, ed. Petra Kuppinger, 1–21. New York: Palgrave Macmillan.

Kuppinger, Petra. 2021c. Sustainability and Small Businesses in Stuttgart, Germany. In *Emergent Spaces*, ed. Petra Kuppinger, 207–226. New York: Palgrave Macmillan.

Kuppinger, Petra. 2023. Incidental Sustainability: Notes from a Thrift Store in Germany. *Critique of Anthropology* 43 (3): 269–288.

Larsen, Frederick. 2019. Valuation in Action: Ethnography of an American Thrift Store. *Business History* 61 (1): 155–171.

Lee, Gini. 2018. Public Space for Changing Times: Reuse Strategies in Transforming the "Wastelands" of Cities. InSubverting Consumerism, ed. R. Crocker and K. Chiveralls, 125–143. London: Routledge.

Martínez, Francisco, and Patrick Laviolette, eds. 2019. *Repair, Brokenness, Breakthrough. Ethnographic Responses.* New York: Berghahn.

Minter, Adam. 2019. *Second-Hand: Travels in the New Global Garage Sale.* New York: Bloomsbury.

Murphy, Fiona. 2017. Austerity Ireland, the New Thrift Culture and Sustainable Consumption. *Journal of Business Anthropology* 6 (2): 158–174.

Schmid, Bendikt. 2019. Repair's Diverse Transformative Geographies: Lesson from a Repair Community in Stuttgart. *Ephemera: Theory and Politics in Organization* 19 (2): 229–251.

Schmid, Bendikt. 2020. *Making Transformative Geographies: Lessons from Stuttgart's Community Economy.* Bielefeld: Transcript.

Steward, Shelly. 2020. What Does That Shirt Mean to You? Thrift-Store Consumption as Cultural Capital. *Journal of Consumer Culture* 20 (4): 457–477.

Strebel, Ignaz, Alain Bovet, and Philippe Sormani, eds. 2019. *Repair Works Ethnographies: Revisiting Breakdown, Relocating Materiality.* Singapore: Palgrave Macmillan.

Ta, Trang X. 2017. A Space for Second-Hand Goods: Trading the Remnants of Material Life in Hong Kong. *Economic Anthropology* 4 (1): 120–131.

Waller, Vivienne, Linda Blackall, and Peter Newton. 2018. Composting as Everyday Alchemy: Producing Compost from Food Scraps in Twenty-First Century Urban Environments. In *Subverting Consumerism*, ed. Robert Crocker and Keri Chiveralls, 186–203. London: Routledge.

PART II

Repair

CHAPTER 7

Ready to Repair: Secondhand Electronics in Tanzania

Samwel Moses Ntapanta

> *August 2020, Essex, United Kingdom:*
> Hello! Hello! Welcome again XX as you all know work determines our dignity. As our normal routine, every Saturday it is a must that we bring to you an auction sale, the largest auction sale. *Makinikia* [concentrate] auction. Today's auction is going to be the biggest one, full of *makinikia*. If you have never participated, give a deposit to A [referring to staff sitting in the company office in Dar es Salaam] and your mobile number will be listed there. If you deposit, you will be allowed to participate in the auctions from today onwards. We will start with chairs, then engineering equipment, then many other toys. We also have several groups for power tools. Today we have a lot of home cinema, a lot of gym equipment. As usual, we thank God for the work of our hands. (Facebook Live post, translated from Swahili by author.)

This auction of used goods, including electronics, happens every Saturday evening on Facebook Live. The auctioneer is located at a depot at a port

S. M. Ntapanta (✉)
Department of Anthropology, School of Culture and Society, Aarhus University, Højbjerg, Denmark
e-mail: ntapanta@cas.au.dk

P. Kuppinger (ed.), *Second-Hand Cultures and Economies of Reuse, Repair, Sharing, and Care*,
https://doi.org/10.1007/978-3-031-99875-1_7

in Essex, United Kingdom. The buyers are in Tanzania, sitting in their homes and participating in the online auction through their devices. Buyers bid through the chat box, and the highest bid wins the round. Although the auctioneer is in the United Kingdom, his team in Tanzania follows and keeps records of each item bought, their prices, and the winner of each bid. At the end of the auction, the auctioneer announces the expected arrival dates of the goods in Dar es Salaam and the procedures for receiving them. That same evening, the goods are packed in containers to clear space for the next round of goods to be auctioned. Every week, this auctioneer sends about two shipping containers to Tanzania.

Introduction

In Tanzania, people continue to buy second-hand goods despite the rise and abundance of new goods. The logic of their purchasing decisions is rooted in social relationships, repair infrastructures, and affordability, which many individuals prioritize over novelty. The availability of second-hand goods in local markets and a ready hand to repair contributes to their continued popularity. These consumer practices are rooted in colonial relations, histories of a people, and, of course, purchasing power. However, the export of used, sometimes waste, goods is deeply embedded in socio, cultural, economic, and political relations between communities from where the objects originate and those that receive them.

Much has been written about used goods in East Africa, for example, clothes (Hansen 2000a, b; Calabrese et al. 2017; Kuiper 2023), shoes (Malefakis 2019, 2015), electronics (Ntapanta 2024; Ntapanta and Medaas 2025), and cars (Raghunath et al. 2023; Grace 2021). This literature explores how certain goods are used, and how they reflect socio-economic conditions. This chapter extends this literature by analyzing the consumer logic and assemblages of use and repair practices and how they contribute to the preference for certain used electronics. I elaborate on details of how second-hand electronics are imported, distributed, used, and repaired in Tanzania. The prominence of the second-hand electronics market in Tanzania only vaguely correlates to patterns of poverty or Tanzanians' purchasing power. Instead, aspects of repair, repairability, and repair infrastructures are crucial in this market. This logic contradicts Hedegaard's broken hammer allegory (Puthussery 2019; Strebel et al. 2018). Consumers in Tanzania think about repair, not when the device breaks, but when they purchase an item. They must know upfront, if the device should break, who, where, and at what price this item can be fixed. Knowing who has the skills to repair things, and at what costs, influences a customer's decision on which brands to buy, whether new or used, or "Chinese" (as they refer to counterfeit devices in Tanzania). These consumer preferences influence repair bricolages, apprenticeship, and the electronic market.

With regard to second-hand electronics exported from the Global North to the Global South, the common argument is that wealthy regions are dumping and exporting harm and pollution to the peripheries (Heacock et al. 2016; Lawhon 2013; Ntapanta 2021; WHO 2021). Studies point to aspects of pollution, which are deeply rooted in (post) colonial relations (Liboiron 2021). However, at a closer look, things are more complex, and electronics are *phamakonic*, both a remedy and poison. When second-hand electronics are exported to Tanzania, hazardous chemicals contained in them are also exported but they are not just ending dumped in some landfill. Instead, upon their arrival in Tanzania, these products enter into new relations and are no longer waste or discarded objects. They emerge as fluid technological devices that weave themselves into relations and practices of reuse and repair that represent affordability and connectivity for many people.

Where Is the e-Waste?

In March 2018, shortly after commencing my PhD program, I went on a field expedition to Tanzania together with colleagues from the Anthropology of Toxicity (AnthroTox) project at the University of Oslo. AnthroTox is an interdisciplinary project that brings together social anthropologists, historians, Science, Technology and Society (STS) scholars, environmental toxicologists, and chemists from Tanzania, Norway, and Germany who conducted research on how environmental, social, and political-economic processes shape flows and impacts of anthropogenic toxicants from electric and electronic devices ("e-waste") in Tanzania and beyond. During our expedition, we visited governmental institutions overseeing waste management and environmental affairs and research organizations. We explored electronic markets and visited landfills in Zanzibar and Dar es Salaam. The team collected data about the substantial quantities of electronic waste that are exported from the Global North to African countries. We further analyzed the impactful visuals presented in Western media reports on illicit waste imports in African countries.

Moreover, we studied the unregulated recycling of electronics. The group vaguely anticipated discovering situations that resembled some of the horrors of the e-waste in landfills in Ghana portrayed in the documentary *Welcome to Sodom*, by the Austrian filmmakers Florian Weigensamer and Christian Krönes (Weigensamer and Krönes 2018; see also Adjei 2014). However, things unfolded differently in our research.

In the bustling Kariakoo and Darajani markets in Dar es Salaam and Unguja, we witnessed sizeable quantities of new and second-hand electronics. We were surprised by the small quantities of electronic garbage disposed in the landfills we surveyed. This limited amount of landfilled material largely consisted of plastic commonly used in electrical devices, such as polystyrene, polypropylene, or acrylonitrile butadiene styrene. There were few indications of informal activities relating to this kind of e-waste. Where then did such e-waste end up? Where was the electronic garbage that we had come to investigate? Where were the items that were no longer functional or could not be fixed and thus hindered the smoothness of life and became matter out of place? Where were places, activities, and labor that dealt with this matter out of place, to use Mary Douglas's analysis of waste (Ntapanta 2024; Douglas 2003)? The AnthroTox team's search for e-waste was guided by these questions. Such thinking, which draws on Douglas's constructivist approach to waste, often dominates waste analysis. It guides explorations of how some observers perceive the arrangement of things in their environment and how objects, which have outlived their assigned purpose or even disturb the established order, become waste. Objects are classified and given a role, purpose, and positionality. Those which challenge or disturb the order or hinder the flow of life are removed from daily life (Bauman 2013; Doherty 2021; Douglas 2003; Reno 2016; Strasser 2000).

When exactly do electronics become e-waste? My field observations from electronics importers in Tanzania illustrate classifications of when or where a device can be/become waste and when it cannot. I followed auctions of used goods, including electronics that take place every Saturday evening on Facebook Live. The devices auctioned there can be classified as waste under the Basel Convention, which considers any electronic discarded by its owner as waste. Returning to the electronics auction in Essex described at the beginning of the chapter: The seller has probably collected the objects that are for sale in second-hand shops, depots, recycling centers, or in streets in the United Kingdom. The devices are discarded objects. They are matter out of place in their previous contexts. Once at the auction and even more so upon arriving in Tanzania, they become matter in place once again. The numerous second-hand shops that the AnthroTox team observed in Kariakoo and Darajani and the online audience at the weekly auction in the United Kingdom demonstrate that waste in the Global North is not waste in Tanzania.

This transformation adds another facet to debates about matter in and out of place and questions about repair and repair cultures. Clearly, in places where people repair, reuse, and repurpose, less waste is produced (Rufus and Lawson 2009; Strasser 2000). People repair a device because they see the value and potential use of the item. It is not waste but remains useful. Across much of the Global North such views on products and related repair cultures are in decline, because of capitalist practices of planned obsolescence. Repair has become too complicated and expensive and buying new is encouraged by corporations and policies. This attitude sharply contrasts with those in places, where people repair and subsequently have a vastly different conceptualization of what is, materially and conceptually, waste.

For example, the analysis of ships being dismantled in Bangladesh demonstrates how "worthless" objects from one place turn into materials for creativity, entrepreneurship, and new consumption patterns in other places (Gregson et al. 2010). When discarded or obsolete ships arrive at Bengal Bay to be taken apart, materials like furniture, motors, and electronic equipment are scavenged, refurbished, and resold. What had been defined as waste, worthless, and ready to be dumped in Bangladesh is transformed into products that have new use and commercial value. Some are reworked to have exceptional aesthetic value that attracts not only low-income groups but also the middle class (ibid). The question of "where" with regard to waste, e-waste and waste leads to an analysis of geographies and landscapes in which "outdated" and broken objects are valued.

How Do Electronics Become Waste

In *Two Treatises of Government*, John Locke laid out fundamental principles on property relations that remain central in today's neoliberal economic ideas. Locke's premises on individual rights to life, liberty, and property ownership have become "virtues" of contemporary societies (Ryan 1988). Individuals have the right to private property if they have labored them from nature. People should exploit (nature) just enough to maximize their happiness and leave what they do not need in the commons for others (Carden 2017). Any product of labor that is not utilized is a failure to humankind; nothing should be left to spoil if others, through their labor, can utilize the value of the product (Wolf 1995). However, contemporary hyper-consumerism neglects the “no waste” provision (Hull 2009). As the right to pursue happiness through consumption is promoted and has become dominant, so has the “right” to discard. Material items or properties serve as a critical element in anthropology for understanding communities. The analysis of property and value draws on ideas that property and property relations are inherently social. Items that have the potential to be valuable possessions may become devoid of worth and remain unclaimed in some situations among groups of individuals, larger social contexts, or historical moments (Ntapanta 2024; Turner 2017). For a better understanding of such process of de-/valuation, we can zoom into micro contexts using anthropological methods to discover how complex and often fluid value and property relations operate.

In recent years, anthropologists have examined the sociality of waste and discussed questions of the politics of waste, relations between waste and humans, and the relations to animals and the planet. Mary Douglas, who famously pioneered theories of waste, stated that waste is a socially constructed category (Douglas 2003). What is waste to one individual, society, place, and time, is not the same elsewhere. Waste is a product of social relations. Despite intensive debates about waste, it remains conceptually and materially contested (Strasser 2000; Moore 2012). Scholars have described waste as disposable parts of our lives (Lawhon 2013), objects that hinder, disturb, or prevent the smooth functioning of daily life, or “essentially disorder, matter out of place” (Douglas 2003). Waste refers to objects of no immediate value, the used-up, dirty, torn, or undesirable (Anderson 1995; Reno 2009).

Scholars in waste and discards studies examine the sociality, social classification, and politics of waste and many point to the complex and fluctuating relationship between waste and value. De-/valuation of an object remains contextual. Thomas Hylland Eriksen and Elisabeth Schober (2017) argue that an “anthropology of waste must be seen as an anthropology of value.” Michael Thomson argues that objects are not permanently in either the categories of waste or value. An object might decrease in value and eventually reach zero value. When an object reaches zero or lower value, it can become rubbish or waste at that moment and place. This, however, can change. The object might remain in a valueless limbo until hitting a social environment that will exhume its value (Thompson 1981).

Objects can travel back and forth between realms of commodity and waste. They can be defined as waste in one place and time, and then accrue value elsewhere or at another time (assuming that they are kept or stored somewhere and not landfilled or incinerated). Thus, objects are not categorically or intrinsically waste, but social contexts define and determine an object's value. Devaluation is a social process that is embedded in a concrete cultural, political, and historical context. Such processes can change along with other social transformations. Over time an object may considerably gain value (even regardless of its state of repair), sometimes even more than the previous value (Thompson 1981). Property, value, and waste are linked to specific contexts. While property and value might exist separately from waste, we can only analyze waste by considering issues of value and ownership. An object may or may not be the property of somebody or some entity, but there must be such some social context for an object to be classified as waste. When faced with waste, the initial question is: to whom does this waste belong, or who is responsible for its disposal? An object is classified as waste because it exists in a social context (Ntapanta 2024). A tree that fell over in a forest is not seen as waste.

When an entity (person, organization, company) chooses to relinquish an object and discards it, they forfeit their inherent ownership rights. After a municipal garbage truck collects discarded items, the entity that previously possessed the object transferred ownership rights to the municipality or garbage company. Landfills or incinerators have their own rules that govern access to their facilities and waste materials. This is even clearer in recycling centers that have acquired property rights over the collected materials that they are now free to sell to others. Under the Lockean no-waste provision, a dumpster diver who retrieves an item from a dumping site would acquire a moral entitlement to that item, irrespective of the legal context. They could exploit the inherent value for their benefit. (The legal reality in many contexts differs from this premise and people or more specifically supermarkets "own" their garbage so long as it is on their premises). The auctioneers in Essex similarly act upon the Lockean no waste or shared waste idea. Regardless of where they sourced old electronics, they understand their value and very crucially the items' revaluation in Tanzania. Upon arrival in Tanzania, defunct electronics are given a second, or perhaps later even a third of fourth life by way of repair and care. A device's lifespan is extended, and it is dramatically revalued (Santos

2020). The "revived" object transcends the category of waste and might look forward to many more years of existence as the valued property of one or more people or households.

The documentary *The Garbage Smugglers*, aired in 2019 by Norwegian TV, Norsk rikskringkasting (NRK), shows how in Norway some electronics are stolen from return centers, where people can drop off their electronic discards and they are then sent to recycling facilities. The stolen items, however, are shipped for further use to West Africa. The smugglers are shown in the documentary as they pick electronic devices, pack them in used cars, and send them on their journey to West Africa (NRK 2020). The documentary does not show much of what happens after the electronics arrive at their destination. It does not engage processes of repair and care. Instead, it quickly moves to introduce informal recycling centers in West Africa where items are taken apart for their materials. The film connects Norwegian recycling center to those in West Africa and largely ignores the fact that many devices will be carefully repaired and used for many years to come. A product might take many years and countless trips to repair shops before a device finally ends up in an informal recycling center. Regardless of an item's concrete journey, Norwegians who drop their electronics in recycling bins regard them as waste. The "smugglers" might indeed be doing the device, the environment, and future owners a huge favor by reinserting things into loops of repair and use. In the process, the devices experience a considerable increase in value.

Second-Hand and Repair: A Different Type of Consumption

I vividly remember the evening of July 16, 1994. After dinner, which we ate usually at 8 pm, I sat with my father, listening to the news on our Panasonic radio. The World Cup was coming to the finals. Brazil was scheduled to play Italy early on the next morning. This was the first World Cup held in the United States of America. Nigeria participated for the first time in the tournament, and so did the newly unified Germany and Russia (following the collapse of the Soviet Union). For many Tanzanians, it was their first time watching the World Cup on television following President Mwinyi's *ruksa* (liberalization policies) that marked the end of the *ujamaa* (socialist) era and opened Tanzania to more foreign products. New market-oriented policies made it easy for individuals to own a TV set.

Liberalization and neoliberal policies dominated this historical moment in Tanzania. One year later in 1995, Tanzania held its first multi-party elections.

On that July day in 1994, my father had made sure the radio had new batteries to avoid missing the action. Several hours before the kick-off, he warned me I would not be awake for the match. He asked who I predicted would win the World Cup. Until today, I still wonder about my answer. I had many heroes in the Brazilian team, but I vouched for Italy. The champion was eventually determined by a penalty shoot-out, which was the first time ever that a team won the World Cup that way. When I woke up the next day, my father delivered my defeat. Robert Baggio had missed the last penalty and Italy lost the match. However, what made my day and quickly made me forget my defeat was that my father told me that the next World Cup in France in 1998, we would watch on our own TV in our own sitting room (Ntapanta 2024).

In reality, it took fourteen years for us to acquire our first TV set. By 2007 when I was in my second year at university, my father had raised enough funds to buy a TV. He traveled by bus from Tabora to Dar es Salaam to buy a TV, a satellite dish, and a power generator. He had to buy the generator because there was no electricity in our hometown. As the eldest son studying at the university, I had to accompany my father on this shopping trip to the Kariakoo market in Dar es Salam. I asked my father before our travels, what exactly he planned to purchase. He explained that he wanted a used Hitachi TV, a new generator, and a satellite dish. There were other brands at second-hand TV shops, like Panasonic and Sharp. Some were cheaper than the Hitachi he wanted. However, he insisted on buying the Hitachi.

My father was determined to get a Hitachi TV because of the availability of skilled repairers, famously known as *fundi*, in our area, who specialized in Hitachi TVs. A *fundi* has vast knowledge and expertise in the creation, restoration, or upkeep of various objects. Swahili lacks distinct terms for individuals who create and individuals who repair. Thus, *fundi* functions as a general term that when followed by a specific noun denotes the particular type of *fundi*. For example, *fundi majiko* is cookstoves crafter, *fundi magari* refers to a mechanic, *fundi simu* is mobile repairer, *fundi cherehani* denotes a tailor, *fundi upenzi* is a mason, and *fundi friji* describes a refrigerator repairer. I have researched the term's origin, but there is no clear documentation of where and how it came to be an (umbrella) noun for makers, tinkerers, or repairers. The only connection I could locate was the Zulu, Nguni, and Xhosa word *umfindisi*, which means a teacher, tutor, priest, guru, or expert (Paton and Soboil 2002). There could be a connection between the Swahili word and the Nguni meaning. In the 1880s, Nguni-speaking people fled in the Mfecane war in southern Africa from the ruthless leadership of Xhaka Zulu and migrated north. Some eventually settled in Tanzania (Thompson 1981).

The availability of a *fundi* was a decisive element in my father's decision to purchase a Hitachi TV. He was apprehensive about other brands that were not as reputable and well known, because it would be difficult to locate a skilled repairer if something went wrong with the TV. Brands are important not only for their own sake and reputation but because of the local availability of skilled repair people. Not surprisingly, in the auction described above, the auctioneer had to visually display the brand of an item and verbally communicate this brand to potential buyers. Much like my father's preference for a Hitachi TV, Tanzanians largely prefer Japanese cars, and here especially Toyota models. Local mechanics excel in the repair of Japanese and Toyota cars. Because of their preference for these cars, however, there is also an abundance of available spare parts for these favored models.

For most consumers in Tanzania when they purchase large consumer goods like electronics, several factors determine which device, new or used, they will buy. Consumers think of affordability, durability, repairability, available infrastructures to repair, and finally status. In certain instances, they might forgo affordability and pay even higher prices for a device they know can be cheaply repaired. A device's durability and availability of a *fundi* who knows how to repair that particular object are crucial aspects in decision making. Of course, people think of devices that are "prestigious" and can elevate their status in society. However, status and price are secondary criteria for most consumers.

There are many new devices available in Kariakoo, and some are cheaper than second-hand models. However, many consumers decide to pay more for a used appliance. The majority of new electronics in the Kariakoo market are made in China. In Tanzania, the term *mchina* devices refers to an item that is cheap, low quality, and difficult to repair. "Original" means good quality and refers to devices of the type that are also sold in the Global North. When shopping in Kariakoo, storeowners or clerk will usually ask whether the customer desires a *mchina* or original product. Originals tend to be more expensive than used and *mchina*. At the same time, high quality and well-known used devices have a reputation of durability. It has been used, but can be (easily) repaired. Consumers tend to choose used devices, because they are durable. They opt for devices and brands that repairers know well. Consumers trust this experience.

Alex Malefikis's observations on *mitumba* (used) clothes and shoes in Dar es Salaam points in the same direction. *Mitumba*, especially from prestigious brands that customers can buy at lower prices than some new pieces, feeds the appetite for conspicuous consumption among the rising middle and elite class. Buying brand name *mitumba* guarantees the quality and authenticity of the product, as being the same as used in the West. It is original and not a counterfeit from China (Malefakis 2015). As my father reasoned about his Hitachi TV, consumers of fashion and clothing anticipate that an object will wear out or break at some point. My father insisted that the TV needed to be used, which means it came from the European market. If such an item broke, it could be repaired without spending massive resources. The new owner only needed to know the appropriate *fundi* who swiftly and expertly could carry out the necessary repair work. People avoid buying brands that are known for being difficult to repair and expensive, for example, Apple products. However, these brands also project a particular social status. Owning an Apple product or a European car confers elite status. At the same time, these products are more expensive to repair. Owning such products requires more income to maintain and repair. Only a small and select group of *mafundi* (plural of *fundi*) know how to repair those brands, which makes repairing expensive and owning them a sign of wealth. Many *mafundi* do not want to invest time and resources in repairing these products as only small numbers of people own such brands, and spare parts are hard to procure.

Conclusion

Consumers who search for products in the second-hand markets of Dar es Salaam or Zanzibar employ criteria such as durability and repairability in general and the availability of concrete repair people in their own towns and surrounding to guide their search. These factors have become almost obsolete in the Global North. Devices are routinely designed to have limited lifespans and are increasingly difficult to repair. Governments have devised policies that hinder third-party repair. Facing such obstacles to repair, consumers opt to buy new devices when breakdowns happen. In turn, the many such broken devices will be collected by used goods dealers and exported to countries, like Tanzania, where a new life journey for the device will begin. One day, of course, the devices will also resist further repair and end up in informal recycling centers (see Ntapanta 2024). Yet, this day might be years of even decades away from the item's first purchase by a Tanzanian consumer. Even upon their final demise when devices arrive at a local recycling center, valorization continues, and the items are not really waste. At informal collection centers, repairers extract and gather spare parts for fixing yet other devices. Urban miners extract copper and other metals that return to production cycles. Only the plastics seem to become waste (what the research team observed at the landfills), all else is somehow reused, repurposed, or recycled.

Regardless of techno-capitalists' planned obsolescence and infrastructures that aim to delink repair from consumption, knowledgeable traders in the UK, their partners, and repair experts in Tanzania realize value what others have designated as garbage. Upon leaving the UK, electronic devices are revalued and put on a trajectory of many years of further use in Tanzania. Consumers there realize this value when they acquire the repaired devices. With a *fundi* at hand, consumers for many year to come are able to outwit planned obsolescence. With repair and care, they postpone the end-of-life moment of their device (Ntapanta 2024). For my father and other consumers in Tanzania, repair is embedded in their thinking about consumption. They will not buy a device without knowing a repairer or that the item can be repaired in general. If, after years of use, devices have reached the status of discards in Tanzania, they enter yet another cycle of revaluation when they are mined for parts and raw materials for yet another round of diverse uses. In the end very little is ever wasted.

References

Adjei, Asare. 2014. Life in Sodom and Gomorrah: The world's largest digital dump via @gdnglobaldevpro. Retrieved 22 January 2021, http://www.theguardian.com/global-development-professionalsnetwork/2014/apr/29/agbogbloshie-accra-ghana-largest-ewaste-dump

Anderson, Warwick. 1995. Excremental Colonialism: Public Health and the Poetics of Pollution. *Critical Inquiry* 21 (3): 640–669. https://doi.org/10.1086/448767.

Bauman, Zygmunt. 2013. *Wasted Lives: Modernity and Its Outcasts*. Oxford: John Wiley & Sons.

Calabrese, Linda, Neil Balchin, and Maximiliano Mendez-Parra. 2017. The Phase-Out of Second-Hand Clothing Imports: What Impact for Tanzania?

Carden, Art. 2017. The Creation of Knowledge in Society: Waste Defined by Property and Exchange. *SSRN Electronic Journal*. https://www.academia.edu/34066255/THE_CREATION_OF_KNOWLEDGE_IN_SOCIETY_Waste_Defined_by_Property_and_Exchange.

Douglas, Mary. 2003. *Purity and Danger: An Analysis of Concepts of Pollution and Taboo*. Psychology Press. New York: Routledge.

Doherty, Jacob. (2021). *Waste Worlds: Inhabiting Kampala's Infrastructures of Disposability* (Vol. 6). University of California Press

Eriksen, Thomas Hylland, and Elisabeth Schober. 2017. Waste and the Superfluous: An Introduction. *Social Anthropology* 25 (3): 282–287. https://doi.org/10.1111/1469-8676.12422.

Grace, Joshua. 2021. *African Motors: Technology, Gender, and the History of Development*. Durham: Duke University Press.

Gregson, N., M. Crang, F. Ahamed, N. Akhter, and R. Ferdous. 2010. Following Things of Rubbish Value: End-of-Life Ships, "Chock-Chocky" Furniture and the Bangladeshi Middle Class Consumer. *Geoforum* 41 (6): 846–854. https://doi.org/10.1016/j.geoforum.2010.05.007.

Hansen, Karen Tranberg. 2000a. Other People's Clothes? The International Second-Hand Clothing Trade and Dress Practices in Zambia. *Fashion Theory* 4 (3): 245–274. https://doi.org/10.2752/136270400778995499.

Hansen, Karen Tranberg. 2000b. *Salaula: The World of Secondhand Clothing and Zambia*. Chicago: University of Chicago Press.

Heacock, Michelle, Carol Bain Kelly, Kwadwo Ansong Asante, Linda S. Birnbaum, Åke Lennart Bergman, Marie-Noel Bruné, and Irena Buka. 2016. E-Waste and Harm to Vulnerable Populations: A Growing Global Problem. *Environmental Health Perspectives* 124 (5): 550–555. https://doi.org/10.1289/ehp.1509699.

Hull, Gordon. 2009. Clearing the Rubbish: Locke, the Waste Proviso, and the Moral Justification of Intellectual Property. *Public Affairs Quarterly* 23 (1): 67–93. https://www.jstor.org/stable/40441517.

Kuiper, Gerda. 2023. The Second-Hand Clothing Trade and Moral Economic Contestations over (Re)Distribution in Tanzania. *Journal of Contemporary African Studies* 2023:1–18. https://doi.org/10.1080/02589001.2023.2193366.

Lawhon, Mary. 2013. Dumping Ground or Country-in-Transition? Discourses of E-Waste in South Africa. *Environment and Planning. C, Government & Policy* 31 (4): 700–715. https://doi.org/10.1068/c1254.

Liboiron, Max. 2021. *Pollution is Colonialism*. Durham: Duke University Press.

Malefakis, Alexis. 2015. Beyond Informal Economy: Street Vending as a Culturally Creative Practice in Dar es Salaam, Tanzania. In *RC21 International Conference on "The Ideal City: Between Myth and Reality, Representations, Policies, Contradictions and Challenges"*. August.

Malefakis, Alexis. 2019. Creating a Market Where There Is None: Spatial Practices of Street Vendors in Dar es Salaam, Tanzania. *City & Society* 31 (1): 119–136.

Moore, Sarah A. 2012. Garbage Matters: Concepts in New Geographies of Waste. *Progress in Human Geography* 36 (6): 780–799. https://doi.org/10.1177/0309132512437077.

NRK, dir. 2020. *The Waste Smugglers*. https://tv.nrk.no/serie/brennpunkt/ekstramateriale/MDDP11001519.

Ntapanta, Samwel Moses. 2021. "Lifescaping" Toxicants: Locating and Living with e-Waste in Tanzania. *Anthropology Today* 37 (4): 7–10. https://doi.org/10.1111/1467-8322.12663.

Ntapanta, Samwel Moses. 2024 Forthcoming. *Gathering e-Waste: Labour, Value and Toxicity in Tanzania*. New York: Lexington Books & Bloomsbury.

Ntapanta, Samwel Moses, and Christian Peter Medaas. 2025. Who cares, and for what? Exploring the contents and extents of care as an analytical concept in ethnographies of electronics repair. *Human Organization 84* (2): 182–190.

Paton, Alan, and Maggie Soboil. 2002. *Cry, the Beloved Country*. Sound Recording. Unabridged. Prince Frederick, MD: Recorded Books, Inc.

Puthussery, Honeylal. 2019. Ways of Being of Equipment: A Heideggerian Enquiry into Design Process. *Tattva Journal of Philosophy* 11 (1): 31–52.

Raghunath, Ananditha, Laurel Krovetz, Hosea Mpogole, Henry Mulisa, Brian Dillon, and Richard Anderson. 2023. From Grasshoppers to Secondhand Cars: Understanding the Smartphone-Enabled Marketplace in Peri-Urban Tanzania. Proceedings of the 2023 CHI Conference on Human Factors in Computing Systems 1–13.

Reno, Joshua. 2009. Your Trash is Someone's Treasure: The Politics of Value at a Michigan Landfill. *Journal of Material Culture* 14 (1): 29–46. https://doi.org/10.1177/1359183508100007.

Joshua O. Reno. 2016. *Waste away: Working and living with a North American landfill*. University of California Press

Rufus, Anneli, and Kristan Lawson. 2009. *The Scavengers' Manifesto*. New York: TarcherPerigee.

Ryan, Alan. 1988. *Property*. Minneapolis: University of Minnesota Press.

Santos, Janine Patricia. 2020. Turn of the Screw: Hacking as Care and Repair in Lomé's Marché Noire Dékon. *Paper present at the ASA Annual Conference 2020*. https://africanstudies.org/annual-meetings-asa/2020-call-for-panelists/geopoetics-of-care-and-repair-in-the-postcolony/.

Strasser, Susan. 2000. *Waste and Want: A Social History of Trash*. New York: Henry Holt and Company.

Strebel, Ignaz, Alain Bovet, and Philippe Sormani, eds. 2018. *Repair Work Ethnographies: Revisiting Breakdown, Relocating Materiality*. Singapore: Palgrave Macmillan.

Thompson, T. J. 1981. The Origins, Migration and Settlement in the Northern Ngoni. *The Society of Malawi Journal* 34 (1): 6–35. https://www.jstor.org/stable/29778451.

Turner, Bertram. 2017. The Anthropology of Property. In *Comparative Property Law*, ed. Michele Graziadei and Lionel Smith, 26–48. Edward Elgar Publishing.

Weigensamer, Florian, and Christian Krönes. 2018. Welcome to Sodom. Documentary Film. Vienna: Blackbox Film & Medienproduktion.

WHO. 2021. Soaring E-Waste Affects the Health of Millions of Children, WHO Warns. WHO. 15 June 2021. https://www.who.int/news/item/15-06-2021-soaring-e-waste-affects-the-health-of-millions-of-children-who-warns.

Wolf, Clark. 1995. Contemporary Property Rights, Lockean Provisos, and the Interests of Future Generations. *Ethics* 105 (4): 791–818. http://www.jstor.org/stable/2382112.

PART III

Reusing and Circulating Clothes and Other Goods and Materials

CHAPTER 8

"This Is Really *Genbrug*!" International Mothers' Consumption of Second-Hand Children's Goods in Copenhagen, Denmark

Erika Kuever

Introduction

Becoming a parent entails learning not just about the nature and needs of your children, but about the material world that surrounds infancy and childhood: the objects that allow them to exist comfortably in the world, and eventually to explore it. The learning curve can be steep for any parent, but especially for those who are not native to the context in which they are raising their children. In Copenhagen, Denmark, where winters are long and cold, but children are expected to spend much of their day outdoors, international parents must learn the names of an array of different garments and accessories, how they function, and where to find them. Most learn quickly about the vast network of second-hand shops, flea markets, and websites devoted to clothes, toys, and furniture for children, and

E. Kuever (✉)
Department of Business and Management, University of Southern Denmark, Odense, Denmark
e-mail: erku@sam.sdu.dk

P. Kuppinger (ed.), *Second-Hand Cultures and Economies of Reuse, Repair, Sharing, and Care*,
https://doi.org/10.1007/978-3-031-99875-1_8

begin to explore them. Although initially motivated by the opportunity to save money, international parents'—particularly mothers—engagements with the materiality of children's goods leads, in most cases, to a preference for second-hand products over new. Second-hand goods, I argue here, have affordances that are particularly well-suited to the needs of both children and parents. Bohlin refers to second-hand objects as "things-in-motion, set on a journey," and it is this trajectory of use that enables engagement, builds community, and allows international mothers to relax and let their children play freely, an ethic that aligns with their experience of parenting in Denmark (2019, 7).

My interest in international mothers' engagement in second-hand economies arose through my participation in them. Originally from the U.S., I had lived in Denmark for six years by the time my daughter was born in 2019, and while I had acclimated to the cold weather and car-free lifestyle, I had no idea how to clothe or transport an infant. I began furtively peeking at the sleeping babies being wheeled around my neighborhood in purpose-built strollers, wondering how the parents knew their children were warm enough. Once my daughter was born I received some advice from the *sundhedsplejerske* (health nurse) who visited our home half a dozen times in the first few months of her life, but it wasn't until my mother's group started meeting that I found a community with answers to my questions. In Denmark every municipality invites mothers to join a group randomly formed on the basis of their children's dates of birth. Fathers may join their own groups but rarely do so. My mother's group was international and had a total of eleven members who had lived in Denmark between one and sixteen years and were originally from France, Italy, Spain, Poland, Taiwan, Scotland, Belgium, Jordan, Finland, and Canada. When we met for the first time most of our babies were around three months old.

One mother arrived at the meeting quite late, struggling with the wrap-style carrier her daughter was sleeping in. A mother from France said it was important to find the style that worked for you and offered the newly arrived mother a different type her neighbor had handed down to her the next time we met. Another suggested visiting the nearby children's charity shop Mødrehjælpen which had a stack of second-hand carriers in many different styles that she could experiment with cheaply. In this first meeting and subsequent ones I was struck by how many of the mothers spoke of buying things for their children second-hand as well as how eager they were to share goods with each other. Although I had grown up shopping in thrift stores, I had assumed my partner and I would be buying most of our daughter's things new. But among the mothers in my group, buying new was the exception rather than the norm On several occasions I heard a mother seeming to justify having bought something new, for instance explaining that she had a specific product in mind and couldn't find it second-hand. It was surprising to me that these affluent mothers—many of whom came from cultures where second-hand goods, especially for children, are stigmatized or seen as unhygienic, were so enthusiastic about them here.

My decision to pursue this research topic also stemmed from my access to this community of mothers. Danish mother's groups have previously been utilized in research on consumption of baby clothes (Andersen et al. 2007), but studies of the consumption of second-hand children's goods by mothers have typically been based at physical sites of consumption such as markets (Clarke 2000; Gregson and Crewe 1998, 2003; Waight 2017). All the members of my mother's group had purchased second-hand goods for their children at some point, but some had primarily done so online, or relied mostly on hand-me-downs. Focusing on this random group of mothers allowed me to capture a broader range of engagements with second-hand goods. I focused on mothers because they were the community I had access to, but also because they had primary responsibility within their families for acquisition, organization and divestment of their children's goods. I interviewed each of the mothers when their children were between 16 and 24 months old and carried out a second round of interviews with five of the mothers a year later. I also received 67 responses from a survey on second-hand buying practices administered through a Facebook group for international mothers. Participant observation was a vital source of data and helped direct my inquiries, although my positionality as a member of this group was sometimes a hindrance as well as a help and required rigorous reflexivity.

Genbrug: Second-Hand Shopping in Denmark

Gregson and Crewe have observed a generational change in attitudes toward the purchase of second-hand clothes for children in the U.K., which they attribute in part to the perception by consumers that the prices for these items are exorbitant, even immoral. They argue that second-hand shopping for children's clothes may be interpreted "as resistant strategies in a broader moral economy of household provisioning" (2003, 125). While new products for children are quite expensive in Denmark, an affluent country with a high cost of living and value-added tax on consumer goods, the mothers I interviewed typically saw these prices as fair, believing that the quality of Danish goods was high. For many of the mothers, buying second-hand was a way to obtain top-quality goods at a more affordable price. It is possible to buy less expensive clothes at international retailers such as H&M or Zara, but the mothers who admitted to doing so inevitably followed by saying they knew they shouldn't because it was 'fast fashion.' Buying second-hand was thus both a strategy for saving money and for avoiding morally tainted wares. All of the mothers also cited the environmental benefits of engagement in second-hand economies, and appreciated living in Denmark, where they perceived the discourses of sustainability and circularity as driving forces in the normalization of shopping second-hand.

My mother's group was dominated by highly educated women, many of whom worked in health technologies or biotech, and had first come to Denmark as students. Others had moved here for jobs in universities or to join Danish partners. While all the mothers in my group had purchased second-hand products for their children, one abandoned this practice a few months in for reasons I will explain later. At the time of our first interviews the other ten mothers estimated that only between one and twenty-five percent of their children's clothes, toys, and furniture were bought new. The majority of these goods were instead obtained from friends or family, at flea markets or second-hand stores, from local recycling stations, or through online sellers or sharing websites.

In the survey, 65 out of 67 respondents reported engaging in second-hand markets for children's clothes, toys, and furniture. The two who did not shop second-hand gave as their reason a concern with hygiene or the general 'ick' factor, but this was almost entirely absent among the mothers I interviewed, many of whom had no problem buying even used swimsuits or underwear for their children (or if they did not do so acknowledged the goods were perfectly safe if washed at a high temperature). The absence of a stigma or concerns about safety or hygiene that Waight (2017) found widespread among mothers in the U.K. can be attributed to the societal acceptance of second-hand goods and their association with environmentalism and extending the use of good quality products.

A concern with sustainability is a pervasive feature of life in contemporary Denmark, where even conservative political parties emphasize their concern for the environment. This is visible in the area of waste and recycling: residential communities in the capital city of Copenhagen offer an array of sorting bins for paper, plastic, metal, compost, small electronics, and textiles. There is an interesting linguistic conflation here: the Danish term for used or second-hand goods, *genbrug* (literally re-use), is the same one that is used for the recycling of materials such as paper or glass. This means that used goods are not immediately associated with thrift, as in the U.S., or with charity, as in the U.K., but with positive environmental outcomes, waste reduction, and circularity.

Messages about sustainable consumption in civic campaigns against fast fashion or marketplace efforts to promote second-hand or upcycled goods also invoke this association. Figure 8.1 shows one such example, a poster

Fig. 8.1 A roadside sign in Copenhagen listing the many occasions for gift-giving during the Christmas holidays and exhorting residents to find gifts at their local recycling centers (photo by author)

from a municipal campaign encouraging residents to find gifts for the December holidays at their local *genbrugsstation*—a recycling center where people can also drop off or pick up anything from books and furniture to tableware and toys. Only about a third of the members of my mothers' group spoke Danish fluently, but all had clearly absorbed the societal valorization of *genbrug*. *Genbrug* stores are everywhere—even in affluent parts of urban areas, flanked by antique stores or hip cafes—and some only carry products for children. The female retiree volunteers who staff these stores promote this discourse of re-use in their own ways. When I arrived one morning carrying a brightly colored musical walker I had purchased there for my daughter some months back, one of the volunteers greeted me chuckling, "I've seen this one a few times before—this is really *genbrug*!"

International Mothers and Provisioning for Children

There is another layer here that relates to the mothers' status as internationals—even though several are permanent residents in Denmark. As foreigners, they are not expected to know the 'right' way to do things—something that came up in several conversations when the mothers spoke of how employees at their children's daycare would give them gentle suggestions on how to clothe their children appropriately for the weather. They didn't feel the weight of a critical gaze as some Finnish mothers in Finland reported to Huopalainen and Satama (2020). Moreover, several were relieved to be free of the judgement of their home societies. This was particularly the case for Lara, one of the Italian mothers, whose parents and sister vocally disapproved of the second-hand clothes she dressed her son in, asking "don't you care about buying him the best? Can't you afford it?" Lara wasn't bothered by this, explaining that she preferred the 'Danish way.' In Italy, she told me, the kids look very nice, but they have to be careful about their clothes, and she wants her son to be carefree in his play.

Many of the mothers embraced this more relaxed 'Danish' style of parenting, implicitly opposing it to the more demanding models they saw in their home countries or practiced by online influencers. This constituted a rejection of the "professionalization of motherhood in late capitalism" where the mother is expected to perform as a childcare expert, an important part of which is purchasing the 'right' kind of commodities (Huopalainen and Satama 2020, 101). This is not to say that mothers' consumption of children's clothes and other products was not shaped by social and political norms and cultural anxieties and desires (Andersen et al. 2007; Clarke 2000; Huopalainen and Satama 2020), only that they were able to align their identities with a model of motherhood that included participation in second-hand economies. Their sense of motherly efficacy was enhanced by the time, knowledge, and labor involved in considering what ought to be bought new and what could be found second-hand, by finding the right product and the right price, by caring for these goods, and by extending their use (Clarke 2000).

Although Denmark has a reputation for high levels of gender equality, and all of the mothers acknowledged that their partners were active and engaged fathers, the work of sourcing and circulating goods for children was care work primarily undertaken by the mothers (see also Clarke 2000; Gregson and Crewe 1998; Huopalainen and Satama 2020; Waight 2014). It is worth noting that in the first year of their children's lives, when children grow most rapidly, most of the mothers were on full-time parental leave, and thus had more opportunities to browse shops, markets, and websites than their partners did. While none of the mothers expressed a desire for their partners to be more involved in provisioning, a few expressed frustration that the work of managing the ever-increasing volume of stuff in their homes not only fell primarily to them but was sometimes exacerbated by the 'hoarding' tendencies of their partners.

Childhood as we understand it today is a fairly recent concept, and changing ideas of childhood are always reflected in clothing for children. Children's clothes are also a means of communicating the class, status, and morality of their families (Rose 2013). In her study of maternal consumption in the nineteenth century, Rose (2013) argues that wealthier families may not have felt the same need to display cleanliness through their children as less affluent families did. A similar dynamic can be seen at work among middle to upper-class women who dress their children in second-hand clothes. In her research in the U.K., Waight finds that "the practice of thrift was a source of pride for the more middle-class mothers," and that unlike their working-class counterparts, "these mothers do not place such high regard on displaying good mothering through the child's dress" (2017, 9). While the mothers in my sample appreciated saving money by buying or receiving second-hand goods for their children, they didn't hesitate to spend more when it mattered—for instance by buying costly new shoes multiple times a year for their unsteady toddlers. Because the mothers were financially able to purchase new commodities, but accepted hand-me-downs or chose to purchase second-hand goods, they were insulated from perceptions that they could not provide the best for their children. In this way participation in second-hand economies was not simply about saving money, but about redirecting resources in a sensible way.

Motivations for Re-Use: Thrift, Narrative, and Conservation

While none of the mothers in my sample used the term in discussing their provisioning strategies, the ethic of thrift clearly played a role in their understanding and enjoyment of these strategies. According to Gregson and Crewe, purchasing second-hand can be a way of "capturing 'value' in the sense of 'the bargain', with its connections to the imperative of thrift" (2003, 11). This imperative has been fruitfully explored by Miller (1998), who has shown that thrift transforms shopping into a moral act and represents "a devotional gift to the future" (102). Miller's treatment of thrift powerfully illuminates the way that participation in second-hand economies for children enables mothers to prioritize the overall well-being of the family by re-capturing resources that would otherwise be devoted to less important things. In my conversations with mothers this idea was expressed through the term 'sense.' It didn't "make sense," I was told over and over again, to spend so much money on something for a child who was just going to muck it up, would grow out of it in a few months, or who would be just as happy wearing something that cost a fraction.

Bardhi and Arnould identify several dimensions of thrift shopping in the U.S., including avoiding waste, careful management of current resources, and extension of products' useful life (2003, 232). The issue of use, which I will expand upon later, is particularly significant with relation to buying for children. Children grow rapidly, especially when they are young, and may only use clothes, toys, or furniture for a short period, or not at all. This surfeit of unneeded—but often affectively loaded—goods itself incites "a desire to re-use rather than waste" (Waight 2017, 3) and spurs their circulation in second-hand and informal economies (see also Clarke 2000; Gregson and Crewe 1998, 2003). Bohlin's work is grounded in the materiality of things and how people relate to their "their ongoing material transformations" (2020, 99). She points out that the physical qualities of second-hand goods "point not just towards the past but the future" (2020, 104). Whether obtained new or second-hand, many of these things are narratively and materially useful.

Things, as we know from Appadurai (1986) and Kopytoff (1986), have social lives and cultural biographies. In her research on clutter, Woodward explains how the expected trajectories of things is an important factor in how objects accumulate in domestic spaces. She describes a free plastic toy—something that often enters the homes of parents of young children—as something that cannot be disposed of instantly even if unwanted, because "things are supposed to have a life before being disposed of as people do not want to be wasteful" (2021, 1224). The historical, symbolic, and geographic biographies of things 'open up' when their use continues past the first cycle (Gregson and Crewe 2003). In her research on end-of-life divestment on behalf of kin, Lovatt identifies a moral imperative to not keep things unnecessarily, but to 'move things along.' She finds that charity shops and other second-hand markets can enable a "successful imagining of a future narrative" for an object (2015, 26).

Divestment is crucial for mothers whose children are at the beginning of their lives yet accumulating large quantities of unneeded things. Furthermore, an inability to part with unnecessary things is a hindrance in what Bohlin describes as "the virtuous ideal of a self-regulating and self-disciplining homeowner, resisting and controlling the world of clutter" (2019, 6). But the potential usefulness of things, a quality Gregson and Crewe call "residual life" (2003, 126), and which Gregson (2007) has described as "surplus," also exerts a pressure on mothers, a pressure to fulfill the expected narrative arcs of goods, to avoid waste, to maximize use. The physicality of things is "intrinsically bound up" in their social lives (Gregson et al. 2007, 688). For Woodward, norms regarding potential usefulness and avoidance of waste are not just brought to bear on objects, but "are an effect of their materiality: the object presents itself as useful—even an unwanted pair of free slippers speak of potential use and rebuke the owner for their lack of use" (2021, 1224). Her research on domestic clutter shows that "through its materiality clutter forces people to engage with moral discourses of wastefulness, usefulness, materialism, and everyday familial norms" (2021, 1215).

Several scholars of second-hand economies in children's goods have seen in this arena a potential for a politics of care for people and the environment but have been struck by the fact that few of those engaged in these economies explain their participation in such terms (Clarke 2000; Gregson and Crewe 2003; Waight 2017). Gregson and Crewe observe that re-use, rather than a politicized practice, is framed more as a practice of conservation "where preserving and/or extending the lives of things has come to matter rather more than thinking about the connections of such practices to the conditions of commodity production" (2003, 9). I argue that the absence of an articulated ethical consumer subjectivity among most of the mothers in my sample is due in large part to their positive, pleasurable experience of acquiring goods for their children.

These mothers are surrounded by discourses of sustainability and circularity that ask ever more of them and that they can never live up to completely. Their choice to participate in second-hand economies of children's goods doesn't require sacrifice but allows them to enjoy the discovery of the unexpected and the pleasures of community (Bardhi and Arnould 2005; Crewe and Gregson 1998; Huopalainen and Satama 2020; Waight 2014). By and large international mothers in Denmark prefer second-hand goods not because they are the more ethical choice, but because they are the more engaging, practical, and affectively rewarding choice. I attribute this in part to the ethic of thrift, but in the second part of this chapter I will also relate it to the themes of ease, engagement, and community that emerged in my interviews. I will also address what I see as a valorization of re-use that stems from mothers' material engagements with used goods for children. I conclude by identifying possible lessons that can be taken from the thriving economy in children's second-hand in Denmark.

Ease

The first and most robust concept around which mothers expressed their preference for second-hand items was ease. As Bohlin argues, used goods invite "more relaxed, intimate and less reverential relations than did new commodities. Less was expected of them...but they also demanded less" (2020, 106). For mothers it was a pleasure to lower their expectations of goods and be easier on themselves. As Spanish mother Isabel admitted:

> I have been disappointed by [new things]. You know, I bought this and then she ended up never wearing it. She didn't like it or grew too fast or it was uncomfortable. And I feel kind of a shame or just a frustration, like oh, that was a bad buy. And I never feel that with *genbrug* you know because it's cheap and it's so easy just to send it back into the system.

Bohlin's work speaks insightfully to the different narrative expectations of new versus used goods. She locates the value of the new commodity "in a conventionalized notion or fantasy that human entanglement has not yet begun" and "all use value of the object remains to be capitalized on" (2020, 110). In the arena of children's goods this means that an unsupervised child with a crayon can ruin a brand-new crib, but only adds charm to a worn but still sturdy crib passed down through generations of children.

Hazel, a mother from Belgium, said she would find it stressful to dress her child in new clothes, stating, "I think it would totally interfere with how we want to live our daily life where I want him to be able to do you know, whatever he wants, and not worry about his clothes at all." Zoe, a French citizen with an international upbringing, chuckled while telling me that the staff at her son's nursery appreciated that they were so casual about his clothes because it meant they never had to worry about finding a matching sock for him. What was being objectified here—in Daniel Miller's (1987) term—was both a style of parenting and a particular idea of childhood: letting kids be kids and letting parents be more than just parents. Second-hand goods were also easy and undemanding because they were considerably cheaper than new products, and because there was a reliable resale market for many of them.

Ease also characterized the experience of shopping for second-hand. The spaces of children's charity shops and flea markets were perceived as asking less of mothers, with none of the aggressive marketing or pushy salespeople one might encounter at a conventional children's store. Zoe spoke of the discomfort she felt when faced with the prospect of buying new: "there is this stress when you go into stores of like, not buying the wrong thing that is exploiting people or made by kids. All of that goes out the window when you're buying secondhand. That really helps." The crucial significance of ease as a factor encouraging second-hand consumption is visible in the testimony of Canadian mother Selena, who felt dealing with second-hand things was too much work and was the only mother in my sample who preferred buying new. Selena had originally planned to use second-hand goods for her daughter and while pregnant had signed up for some 'rotation groups' that facilitated the circulation of children's clothes. After her daughter was born, however, she realized that most of the clothes were either too small, too worn, or out-of-season. Selena felt sorting through and disposing of the unwanted clothes took too much "mental energy" at a time when she had very little and kept only about one bag out of the ten bags she had received.

Engagement

The higher degree of engagement required by second-hand goods was undoubtedly more work, but for most of the mothers it felt both rewarding and like an expression of care—in part because it was a way of diverting resources that could be used elsewhere. Monika, a mother from Poland, spoke about searching for and finding something her son needed, using different networks and marketplaces until she found the best quality at the right price: "I just, you know, feel proud and happy and fulfilled." Engagement, as this quote illustrates, is also about knowledge. Parents—and international parents in particular—must educate themselves about the things their children wear and use but also how they use them. Their clothes may incur specific wear patterns due to the way they play. They will almost certainly get dirtier and require more washing than adult's clothes, which, as many mothers observed, revealed the poor quality of clothes bought from fast fashion retailers such as H&M.

It is not necessary to buy such clothes new to discover how they wear—second-hand stores are a continual source of information about the evolution of objects that circulate. The previous use of an object by another child offered more information about the good than would be available had it been purchased new. Mariam from Jordan told me she was always a little suspicious when she saw an item that appeared to be brand new at a second-hand store or flea market, telling me about a rainsuit she had bought only to discover that the Gore-Tex collar rubbed her son's neck uncomfortably. A Finnish mother, Sari, mentioned the piles of toddler-sized jeans she always saw at her local children's charity shop, telling me she realized "it's because they're cute but kids don't like wearing them—they can't move around in something so tight." On the other hand, there were objects, particularly toys, that appeared to have been used intensively for decades, and this offered evidence of their 'classic' nature and high quality. The increased value of goods that can be re-used many times has also been embraced by brands. Figure 8.2 shows an inside label for a child's snowsuit which leaves space not for one child's name, but three.

A deeper engagement with second-hand goods is necessary because they do not come with the same expectations of material stability as new products, but it also leads to a closer, more intimate relationship with

Fig. 8.2 Name tag for a *flyverdragt* (literally 'flight suit'), a padded, waterproof suit children wear outside in winter. Some tags have room for even more names, implying greater durability (photo by author)

objects, and a keener understanding of their utility. Mothers were attentive to how use could affect the function of a good—what Bohlin calls a "somatic dialogue" (2020, 98). For instance, Isabel showed me a used 'walking bicycle' she had purchased second-hand, indicating a few scuffs and scratches she said she didn't mind, but then pointing at the wheel treads, and explaining that if those had been worn down the bike wouldn't be safe for her daughter to ride. Bohlin finds second-hand items interesting "in the way that they induce particular responses to their obvious 'ongoingness'" (2020, 97). This was clear not only in the way mothers spoke about and interacted with second-hand goods, but in the testimony of the one mother who primarily purchased new goods. Selena thought it was nice to not have to deal with things not working or fitting perfectly well, but Zoe appreciated the opportunities presented by things she or her partner found or received: "it's like oh this could be cool, how are we going to make it work?" For Italian mother Elena obtaining a classic but visibly worn Danish Tripp Trapp highchair was an opportunity to stencil it with a beautiful design.

Community

The sharing or gifting of second-hand goods fostered affective ties in my mother's group, but each mother was also embedded in their own network through which gifts and hand-me-downs circulated. Some received things for their children from faraway friends or family, but the most significant and systematic exchanges typically took place in their local network. These could be one-off deliveries of hand-me-downs of the kind described below, or semi-formal chains of friends or neighbors through which goods for children would move between families. Children's furniture in particular was often lent out for a period of years by families who expected to have more children in the future. Clothes and toys were more likely to simply pass from families with older children to those with younger children. Zoe lived in a building with several families and described the precise path things would take on their way to her son by indicating the different units. Another neighbor was now pregnant, she told me, so she would pass things on to her eventually.

Having been recipients of goods for children, mothers also felt an obligation to give back within their communities. Elena told me she enjoyed sharing children's clothes with her mother friends but also felt it was something she had to do "because I receive a lot of things so I have to give back." While some of the mothers sold some of the bigger items they had bought for their children—for instance cribs or swings—they felt it was inappropriate to sell something they had received for free or bought cheaply at a second-hand boutique. I felt this tension myself on occasion, knowing I could re-sell some LEGOS or a playmat for much more than I paid for it, but also feeling it should return to charity-driven circulation. Monika articulated this when telling me about the 'second-hand boutique' at her local municipal recycling station. She often found bicycles and outdoor play furniture there, things that are always in demand on second-hand shopping sites, but said she felt obliged to return everything she took so it could be used by another child.

Mothers were often 'welcomed' into this broader community of family-based giving and sharing by relative strangers. Many of the mothers, myself included, were given their first hand-me-downs by colleagues or friends of friends. For French mother Julie:

> My experience is as soon as you say you're expecting your first child, suddenly, like, a community builds around you. My workplace is very Danish. At the point when I got pregnant with my first, I had been working there for three years I guess so I didn't have very strong ties with my colleagues. But a girl I didn't know well—she came with two bags of clothes, and another one gave me a bed. And then my neighbor gave us our changing table.

Having access to this world of circulating objects made the mothers feel part of a community that both abstractly and concretely offered support to them. Zoe told me the advice she would give to a new mother was to not worry too much about the things they would need, because "you're gonna discover there's this underground mafia of women that actually hold the world together…and they'll be able to find you stuff."

This feeling of community was also present in more commercial transactions. Several of the mothers in my sample felt a strong affection for their local children's second-hand store, not just because of the volunteers who staffed it and who welcomed our children, but because visits there marked their growth—we would arrive with a bag of things they had grown out of and leave with a bag of things they had grown into. All but one of the women in my mother's group used the app Reshopper, where trades can be arranged for pick up or delivery, and many of us enjoyed pickups as opportunities to peek into other people's homes and see the children who had previously used something. Monika from Poland spoke animatedly about the experience:

> Whenever I pick up an item, I usually meet a nice person on the other side, and that person is super happy seeing me with my son and knowing that this item will get a new life and it's not being thrown away. I often hear, I hope you will enjoy it or I hope you will be happy with it or I hope you can still use it till whenever. And it also makes me happy that we don't have to buy something new. That we're not harming the environment or polluting our planet with overproduction. It's so cool that someone is happy to sell it or get rid of something and I'm happy because this item can now be with us and then maybe later with someone else.

The Valorization of Re-Use

I have argued that ease, engagement, community, and the ethic of thrift are the most important factors behind international mothers' preference for second-hand children's goods. But a broader theme of use and its extension runs through the purchase, circulation, and disposal of these goods. Isabel, for instance, identified environmental concerns and saving money as her primary motivations for buying second-hand, but when asked to elaborate told me "it just feels logical to do it, like, why throw away clothes that are still useful, right? When they can have a second life, third life, fourth life. And why buy new things when there are so many [things] out there already? It just feels right I guess." This valorization of re-use, I argue, comes from mothers' own experiences with objects—their discoveries that things used by other children remain useful to their own, and that things used by their children retain the material potential to continue on a trajectory of use.

I heard re-use valorized in several ways. First, re-use was a way of avoiding waste. All of the mothers avoided throwing things away, and at least three of them would sometimes repair garments or toys to increase the likelihood of their continued use. As Hazel, who sometimes dyed her son's stained clothes before she donated them, told me, "whenever possible I feel like I should pass that on to the next person, you know, deliver it back to the community in as nice of a state as I can." A few of the mothers also expressed a preference for temporary use or sharing over permanent ownership. As Zoe put it "I almost always think of those things as stuff that I'm currently using, and that I would like to pass on if possible."

It was for this reason, she told me, that she rarely wrote her son's name in the spaces designed for it on some of his clothes. I had a similar feeling about my daughter's books, typically leaving the "This book belongs to" space on the cover page blank.

The frequent sorting of children's clothes and toys discussed above was a means to avoid clutter, but also to prompt re-use. Mothers often spoke of wanting to 'keep things moving' or avoid 'holding on' to things that were not in use, almost implying agency on the parts of the objects themselves. Elena erred on the side of circulation when sorting through and getting rid of her children's things: "…maybe I got rid of something they could have worn another time. But who cares? I mean, they did their time, right? It's fine. They get another life." This ability to give something a new life could also be affective, especially when things were passed down to the children of close friends. Several mothers spoke of feeling moved when they saw other children using things that had once belonged to their children. Finally, re-use was valorized when mothers praised the high quality and durability of Danish products. Monika gave the example of the classic Tripp Trapp highchair, noting "you can buy chairs that are almost 20 years old and they are still functioning, it's great. And Danish people or people living in Denmark don't mind buying a chair that is 20 years old and paying 200 kroners for it."

Conclusion

By the time of my second round of interviews—when our children were around three years old—most of the mothers were buying far fewer second-hand clothes for their children. Zoe and Hazel were exceptions. Both environmental activists, they remained committed to buying new only when absolutely necessary, though they admitted that it was getting increasingly difficult and time consuming to find things in larger sizes. Many of the other mothers I spoke to had simply fallen out of the habit of shopping second-hand. Realizing there were fewer things available for bigger children who wore their things longer and harder, they had gotten used to looking for things in online sales, rather than checking Reshopper or their local second-hand children's boutique. It wasn't, they admitted, nearly as fun or satisfying. I argue that the appeal, and even pleasure, of used goods for international mothers in Denmark is related to their specific affordances, allowing mothers to relax and children to play at ease, obtaining information and expressing care through material engagements, building community, and extending the use of objects. Acquiring these goods is also associated with fun, pride, and social engagement, and enables an ethic of thrift.

This research shows that cultural norms that valorize re-use, repair, and sharing can produce new orthodoxies around the value of second-hand goods. Second-hand consumption in Denmark has been de-stigmatized and popularized, and this is not a process that can be exported wholesale to different cultural contexts. However, the thriving market for used children's goods in Denmark does offer some generalizable policy lessons for increasing the consumption of second-hand goods in other domains and/or other parts of the world. First, it is clear that the high price of goods in Denmark—in part due to a high value-added tax—discourages parents from buying new. Second, the association of second-hand practices with sustainability may allow those whose motivation is more financial to frame their actions in a more socially acceptable manner. Finally, if the pleasures of second-hand—at least in the realm of goods for children—are not just about treasure hunting and finding deals, but about engagement, ease, and community, then policymakers and marketers should highlight these affordances in their messaging.

References

Andersen, L. P., E. Sorensen, and M. B. Kjaer. 2007. Not Too Conspicuous, Mothers' Consumption of Baby Clothing. In *European Advances in Consumer Research*, ed. S. Borghini, M. A. McGrath, and C. Otnes, vol. 8, 94–98. Association for Consumer Research.

Appadurai, A., ed. 1986. *The Social Life of Things: Commodities in Cultural Perspective*. Cambridge University Press.

Bardhi, F., and E. J. Arnould. 2003. Thrift Shopping: Combining Utilitarian Thrift and Hedonic Treat Benefits. *Journal of Consumer Behaviour* 4:223–233.

Bohlin, A. (2019) 'It Will Keep Circulating': Loving and Letting Go of Things in Swedish Second-Hand Markets. Worldwide Waste: *Journal of Interdisciplinary Studies* 2(1): 3, 1–11.

Bohlin, A. (2020) The Liveliness of Ordinary Objects: Living with Stuff in the Anthropocene. In Harrison, R. and Sterling, C. (eds.) Deterritorializing the Future: Heritage in, of and after the Anthropocene. London: Open Humanities Press.

Clarke, A. 2000. 'Mother Swapping': The Trafficking of Nearly New Children's Wear. In *Commercial Cultures, Economies, Practices and Spaces*, ed. P. Jackson, M. Lowe, D. Miller, and F. Mort, 85–100. Berg Publishing.

Gregson, N., and L. Crewe. 1998. Dusting Down Second Hand Rose: Gendered Identities and the World of Second-Hand Goods in the Space of the Car Boot Sale. *Gender, Place and Culture* 5 (1): 77–100.

Gregson, N. 2007. *Living with Things: Ridding, Accommodation, Dwelling*. Sean Kingston Publishing.

Gregson, N., and L. Crewe. 2003. *Second-Hand Cultures*. Berg.

Gregson, N., A. Metcalfe, and L. Crewe. 2007. Identity, Mobility, and the Throwaway Society. *Environment and Planning D: Society and Space* 25 (4): 682–700.

Huopalainen, A., and S. Satama. 2020. 'Writing' Aest-ethics on the Child's Body: Developing Maternal Subjectivities Through Clothing Our Children. *Gender, Work and Organization* 27:98–116.

Kopytoff, I. 1986. The Cultural Biography of Things: Commoditization as Process. In *The Social Life of Things*, ed. A. Appadurai, 64–91. Cambridge University Press.

Lovatt, M. 2015. Charity Shops and the Imagined Future of Objects. *Culture Unbound* 7:13–29.

Miller, D. 1987. *Material Culture and Mass Consumption*. Oxford: Blackwell.

Miller, D. 1998. *A Theory of Shopping*. Cornell University Press.

Rose, C. 2013. Maternal Consumption: A View from the Past. *Journal of Consumer Culture* 13 (2): 178–198.

Waight, E. 2014. Second-Hand Consumption Among Middle-Class Mothers in the UK: Thrift, Distinction and Risk. *Families, Relationships and Societies* 3 (1): 159–162.

Waight, E. 2017. Hand-me-down Childrenswear and the Middle-Class Economy of 'Nearly New Sales'. In *Sharing Economies in Times of Crisis: Practices, Politics and Possibilities*, ed. A. Ince and S. M. Hall, 96–109. Routledge.

Woodward, S. 2021. Clutter in Domestic Spaces: Material Vibrancy, and Competing Moralities. *The Sociological Review* 69 (6): 1214–1228.

CHAPTER 9

Frictional Infrastructures: Navigating the Socio-Spatial Landscape of Second-Hand and Waste Tyres in Mega-City Lagos (Nigeria)

Simon Coleman, David Garbin, and Andrew Esiebo

Photography: Andrew Esiebo.

S. Coleman (✉)
Department for the Study of Religion, University of Toronto, Toronto, ON, Canada
e-mail: simon.coleman@utoronto.ca

D. Garbin
School of Social Sciences, University of Kent, Canterbury, UK
e-mail: D.Garbin@kent.ac.uk

A. Esiebo
Purley, UK

P. Kuppinger (ed.), *Second-Hand Cultures and Economies of Reuse, Repair, Sharing, and Care*,
https://doi.org/10.1007/978-3-031-99875-1_9

Setting the Scene in the "Pneuma-City"

In Onike, a busy neighbourhood in Lagos, Nigeria, two men operating a small white truck make a swift stop near the stand of a street tyre repairer, or "vulcanizer" as they are better known. One of the men gathers five tyres that had been piled up on the roadside and proceeds to load them onto the already laden truck, without exchanging a word with the vulcanizer busy fixing the tyre of a *marwal* (tricycle taxi). This urban choreography unfolds in less than a minute. It will be replicated numerous times throughout the day, not only by this specific truck but also by other similar vehicles operating across the city. The tyre collectors, though driving a truck with a small hand-painted inscription LAWMA (Lagos Waste Management Authority), are not employees of the public body but work for one of the many small-scale private operators licensed by LAWMA to collect waste on its behalf. This arrangement provides a microcosm of the increasing deployment of public/private partnerships in much of the Majority World, and is emblematic of more general, global impacts of neoliberal policies on municipal services. It is also indicative of a complex and local context of "waste congestion" (Salvaire 2019, 2) where, for some observers, the "challenge of managing trash [...] acts as potent metaphor for the African 'crisis' writ large" (Fredericks 2018, 23)—pointing to the technical and ethical intricacies of urban management.

Tyre waste, because of its materiality, bulkiness and resilience, is a distinctive form of litter, requiring a specific discard infrastructure. Much of this material follows its own path and value chain: collected from the streets of Lagos (near commercial garages or roadside vulcanizers for instance), it is often disassembled in peripheral dumpsites where its core constituent materials—fabric, rubber, metal—are meticulously segregated for reuse or resale.[1] Tyre Derived Fuel (TDF) obtained from the shredded rubber[2] has become an important alternative energy source, especially in the cement industry, a booming sector in Africa and one that is key to production of the "concrete materiality" of unbounded mega-urbanization (Choplin 2019). Yet not all waste tyres are reintroduced to the industrial value chain as fuel, as old tyres may find new life in the form of repurposed goods, from shoe soles to automobile components (such as cam belts) and even art and furniture pieces. As we shall see, they can also be used more "informally" in numerous creative ways, in particular to fill functional gaps in the city, as parking delineators or manhole covers.

The textual and visual material presented in this chapter draws upon an ethnography as part of the "Pneuma-city" project,[3] through which we examined the urban economies and ecologies associated with both tyre waste and tyre work in urban Nigeria.[4] The project's broader approach conceptualizes tyres as "frictional infrastructure" embodying a dialectical relationship with the urban fabric as they are positioned at the interfaces of roads and vehicles but also in relation to mobile actors—commuters, drivers—and roadside workers such as the vulcanizers. Friction can refer to the literal action of one surface against another, as a tyre rubs against the road, which causes resistance and material loss (including polluting micro-particles) but is also necessary for the mobility of the vehicle resting on its wheels. Tyres are, after all, the only parts of a motorized vehicle in permanent contact with the road—they are the "shoes of the car" as one vulcanizer put it. The imagery of tyres as frictional infrastructure sheds light on the indispensability of roads as conduits of mobility, while the quality of roads, in turn, reflects the significance of tyres in facilitating urban movement while providing a commentary on the physical care (or lack of it) provided by the state or its private contractors.

The material condition of tyres—which, in Nigeria, are often second-hand imports known as *tokunbo*—dictates the necessity of seeking the services of a vulcanizer, thereby intricately linking the maintenance of tyres to the fluidity and rhythms of urban movement. This frictional, reciprocal relationship unveils the intricate dynamics at play, where the state of infrastructure shapes and is shaped by ongoing human activities. In that sense, we find the imagery of friction useful because it refers both to a social scientific metaphor and to the material reality of tyres functioning as mediating "zones of engagement" (Tsing 2005) between different infrastructural, physical, ecological and economic domains in the rapidly urbanizing mega-city, where mass road mobility is crucial to connectivity and livelihood.

Of course, the vital yet taken-for-granted role of tyres has a complex history. The advent of the vulcanization process[5] in the 1830s heralded a transformative era, symbolizing the dawn of the "rubber revolution" (Tully 2011, 40). Rubber, with its newfound durability and versatility, quickly became a prized commodity, playing a role akin to that of coal, oil or steel in fuelling industrialization. Yet, beneath its functional and mundane character lies a darker past tainted by global exploitation, oppression and war—making it, as Tully (2011) coins it, the "Devil's milk" for those who toiled to extract or transform the precious commodity, from

Congolese forced laborers under Belgian King Leopold II, to Amerindians in the late 1900s Brazil and Peru, to Auschwitz prisoners. This crystallized history of rubber, but also the labour relations involved in the latter's transformation, spatial trajectory and value chain, are in Marxist terms "invisibilized" by the commodification process.

Given such history, looking through the medium of the tyre provides us with a powerful opportunity to interrogate the sociality and spatiality of "pneumateriality" though both textual commentary and street photography. To do so, we have divided the rest of this chapter into three parts, which explore the fate and the form of tyres in different though complementary ways. The first interrogates the physical qualities and the material repurposing of tyres in relation to manifold dynamics of their reincarnation, discussing aspects of the politics of value attached to the circulation and transformation of seemingly waste objects. The second takes a visual approach, documenting the presence and impact of relatively unaltered tyres that have become detached from vehicles and are distributed across the urban landscape. This section will provide a glimpse of the visual ethnography and photographic work of Andrew Esiebo,[6] who offers a unique perspective on Lagos, revealing the position, role and trajectory of tyres in the Pneuma-city. The third section focuses on the material sociality of the labour associated with work of repair, focussing on vulcanizers who, in their own words, "keep the city moving."

"A Tyre Never Dies…"

When we examine the full life cycle of tyres in Lagos it becomes evident that their utility extends far beyond their initial function. Besides their reuse as upcycled objects or their transformation into fuel, old tyres in mega-city Lagos can take on a more "organic," informal reincarnation that is connected in new ways to the urban fabric, repurposed for instance as delineators for parking spaces or supports for market stalls. Such "acts of remaking" (Alexander and Reno 2012, 15) provide a kind of reframing: they do not involve recycling or material transformation but are a testament to the sheer durability and perpetual relevance of tyres as articulated by one of our research participants: "In Nigeria, a tyre never dies." His statement gives the lie to the frequently used label "ELT," which stands for *End* of Life Tyre. In environmental terms, of course, this material resilience is both a blessing and a curse, allowing (nonbiodegradable) tyres to cling in sometimes desired, sometimes unwanted ways to the urban fabric, potentially becoming breeding grounds for malarial mosquitos or blocking waterways and drains. One artist whom we interviewed referred in a memorable way to the effortful experience of working with tyre rubber as a medium of expression. He called the latter "very rigorous to manipulate" so that "there are some tyres I tag 'rebel.' Which means that if I try to use it for a particular subject, it is not going to be submissive to me…That's why it's quite different from any other kind of recycle."

The functional ambiguity of tyres emerges further from their characteristics as objects that are intimately connected with daily existence and yet are not liable to become objects of particular affection, given that they retain a certain anonymity in their "biographies" (cf. Kopytoff 1986). While a tyre will show signs of use over time, such marks do not provide it with a personal history that can be linked in memorable or nostalgic ways with its user. Similarly, in contrast to gift and commodity exchange, most tyres can be obtained from previous owners who are likely to be pleased to offload them (compare Reno 2009, 38, drawing on Carrier 1995). As mass-produced items that are not replaced in regular intervals, they do not participate in predictable and/or long-term systems of exchange and circulation that might readily encourage the development of social ties between consumers and sellers (cf. Appelgren and Bohlin 2015). For instance, another artist describes the opportunistic way in which he gathers his materials:

> I have people on social media who message me: "Hey bro, I have these tyres I have gathered, I don't know where to dump them?" People like that I ask them where their location is, if their location is not far from mine, I go to pick up [the tyre(s)]. But if it is far, I sometime let them know that they can give to this person because he or she is closer to you than I am.

Under such circumstances, tyres may even be obtained free. While he also chooses to pay vulcanizers for the opportunity to obtain tyres from them, this artist regards such payments as his own choice—a way of encouraging them to refer to him the next time he needs materials for his art.

This fluid exchange system underscores the complex social (and economic) networks that facilitate the circulation of used tyres within Lagos. Here our analysis of the material and symbolic reappropriation of "end-of-life" tyres points to a nuanced understanding of the politics of value around waste and waste transformations. The plural trajectories of ELTs and the differentiated value and labour regimes identified have led us to adopt a non-reductive lens to capture the complex dynamics within the urban landscape of tyre waste distribution and transformation. Within this landscape there are actors actively engaged in processes of reclamation, transformation and repurposing, their acts of remaking and re-valuation reconstructed in terms of sustainability and circular economy principles, and guided by moral imperatives or aesthetics considerations, such as those of the "tyre artists" we spoke to. This perspective challenges the simplistic view of recycling in the Global South as merely driven by self-sufficiency and necessity (Micklethwaite 2021), highlighting instead shifts in consumption patterns (e.g. growing tastes in upcycled objects) linked to class habitus or the "vernacularisation" (Levitt and Merry 2009) by civil society actors, entrepreneurs, policymakers or NGOs of global environmental agendas.

The work of Jumoke is illustrative of this point. As the creator of a small museum in Ibadan dedicated to waste, sustainability and the circular economy, she learned tyre upcycling techniques from a Lagos-based artist. Reflecting on their material ambiguity, she values tyres' longevity: "I do furniture pieces with tyres, they are very hard to cut and manipulate. But this difficulty means it will last better…So the con is the pro." The items she sells and (advertises online) "are not devalued because they are made of second-hand tyres" as she puts it; instead "what is important is the value you place on the product, the way you talk about them and the value you put on the work." By strategically positioning her upcycled products within narratives of sustainability, durability and artisanal effort, Jumoke engages in a form of value production that elevates the status of upcycled materials, craftmanship and labour beyond purely market-based considerations.

In addition, Jumoke has established a social enterprise through which she organizes educational and training activities on recycling and repurposing tyres, plastics and fabric for young people and local communities in Ibadan. Such activities are sometimes arranged as part of initiatives such as the Global Recycling Day or World Environment Day, and/or supported by funding from global charities and donors. These connections, her engagement with a wider environmental agenda and the importance she places on Sustainable Development Goals (SDGs) suggest the local appropriation and adaptation of global frameworks. We see how local actors can reframe waste materials within new value regimes, blending sustainability with an entrepreneurial ethos alongside aesthetic imperatives.

Tyre upcyclers like Jumoke take great pride in "beautifying waste" and "turning waste into wealth." As Reno (2015, 563) points out, this process "involves an appreciation for the capacities of things to become and not only to contaminate." In other words, such labour defies reduction to "mere" engagement with contamination, embodying a latent potential for creating transformation. Even as waste management assumes a biopolitical dimension (often associated, for instance, with precarious and demeaning activities),[7] it requires a nuanced understanding beyond somewhat static symbolic connotations as "matter out of place" (Douglas 1966). As Jumoke points out:

> Vulcanizers are used to what they are used to: dirty tyres. When we show them the beauty that can come from an old tyre it changes their perspectives....I realize that people have that perception of waste as dirt because they have not been introduced to a better way of using waste.

Mediated by a different form of knowledge, changed "perspectives" can potentially have an impact on the micro-politics of value within the tyre waste-upcycling chain. Jumoke reports:

> Some of the vulcanizer[s] that I collect tyres from, I go with the products that I make the tyres from to show them. And once I go, they increase the price of the tyres I collect from them [laughs]....I have to pay more because I am on a mission, my mission is to create a sustainable world without waste and I know I cannot do it alone.

In contrast, another tyre artist we interviewed adopts a more "covert" approach, reflecting a different facet of these politics of value. While tipping vulcanizers for used tyres, he avoids fully disclosing the nature of his work to prevent them from demanding more:

> I work with some of these vulcanizers at the junction. Most of them will just be piling tyres up for me, I would pick them up and I would give them a tip, but I don't tell them what I do with it. Because they don't have anything to do with it—but they are always curious. I give them different stories. They don't know the usefulness or they wish they could just get their own bite.

While not operating outside of market relations, this practice indicates a strategic navigation of the social and economic dynamics within the tyre upcycling ecosystem. The artist's reluctance to discuss his work points to a delicate balance of maintaining boundaries and control over the value generated from upcycled materials while managing his social and economic distance to vulcanizers as (direct) local suppliers. In his eyes, the value is (or should be) generated by the craft and artistic expertise/experience crystalized and realized in the transformative process (rather than by the knowledge about this potential transformation). Regardless of how these politics of value operate and how knowledge circulates within these systems of division of labour and exchange, (pneuma)-waste thus assumes a multivalent and productive (after)life beyond its symbolic bordering function. Another Lagos-based tyre artist we interviewed provided an unconscious refutation of Douglas's famous maxim, in emphasizing the resilience and adaptability of tyres:

> There is still something you can use them for, if you are not using them because they are very bad....So there is no tyre that can be completely out of place....So, no matter how small or bad a tyre is, there is someone out there than can still reuse it, it might be bad for all but it is good for someone.

However, we may also consider the extent to which, in the context of tyre artistic craft and upcycling, dirt-as-"out-of-placedness" also becomes a pivotal category of valuation. The low economic exchange value of old tyres reflects their lost functionality and exclusion from craft and artistic spheres when they are piled up next to vulcanizers' roadside stands. After all, tyres carry the dirt of the city and roads throughout their life: they are kept outside of the domestic space, like dirty shoes left at the doorstep. Upcyclers purchase these tyres cheaply from vulcanizers, and through their beautifying labour they not only cleanse but revalue these materials, moving them from the periphery of waste into the centre of aesthetic and economic appreciation. In some cases, these old tyres even reincarnate into furniture pieces, entering the domestic space for the first time. The implicit perception of dirt, therefore, both diminishes and potentiates value, depending on its relational context and the transformative labour applied.

(In)visibility and Material (Re)vitality in the Urban Landscape

Our discussions of dirt as both symbol and substance have introduced questions relating to the shape-shifting qualities of tyres as they move through different regimes of both value and use. We also see how tyres (as both discrete objects, and as polluting micro-particles) epitomize the seen as well as the unseen. When affixed to vehicles and operating as expected, they seamlessly blend into an overall kinetic infrastructure, facilitating circulation and mobility—a testament to their essential role in keeping automobile-centric societies functioning. That said, in contemporary megacity Lagos tyres that have become detached from motorized vehicles are observable in many—almost ubiquitous—ways. Abandoned on the roadside, along riverbanks, in canals, wastelands or other interstices of the city, they become waste matter that, given their resilience, embodies the idea of what we would call "impossible ruination." In documenting this ubiquity of tyre waste in Lagos, Andrew Esiebo presents vignettes of the everyday city, contrasting the stasis of tyre waste with the incessant flows of vehicles—including the iconic *danfo* mini-buses (Photo 9.1).

Photo 9.1

This photo frames the multilayered reality of the mega-city through signifiers left at its material surface. In the image, these urban—informal—signifiers are ambivalent or elusive in the sense that their semantic promise is never certain to be fulfilled. The possibility of *connecting* to financial opportunities (including in the form of loans) as literally written on walls reflects the idea of Lagos as a place of individual progress and betterment. Such writing teeters between advertising and graffiti, playing into cultural politics of both dirt and possibility (compare Thurlow 2022, 389). But in a city teeming with aspirations, the line between genuine opportunities and predatory schemes is often blurred, leaving individuals open to possible exploitation and "419" scams.[8] The gritty wall in the midst of incessant urban flows indicated by the blurred vehicles in the background becomes a canvas onto which the hopes but also vulnerabilities of Lagosians—in particular, new Lagosians—are projected.

In Photo 9.2, the abandoned tyre nearly disappears from view, blending seamlessly with a patch of *natura urbana* (Gandy 2024), the slow shutter speed capturing another striking contrast between the stillness of the tyre and the perpetual motion of urban vehicles. Here the juxtaposition and

Photo 9.2

conjoining of two walls tells its own story, though this time a non-verbal one, with the vertical extension and added spikes on top of the wall on the right reflecting an increasing focus on securitization in urban Nigeria (Smith 2022).

Commenting on Photo 9.2, Andrew notes that,

> what struck my attention was this dry green space, also this fence, which has these spikes—a reflection of our society.... Even in terms of the architecture...you see this inconsistency of the bricks....When there was less insecurity, fences were not this high. But because of the insecurity now, more people will have to make their fence higher and even put spikes on top of it. Those are the kind of things that when I'm looking, I'm thinking and relating to the story—why this and that is there. I play with motion because for me that is aesthetically appealing but also, it's to find a way not to make the motorbike the centre of attraction but to lead you to the subject of tyre[s].

In Photo 9.3, taken during the rainy season, tyres again seem organically integrated into the urban fabric, but here they are put to use in enabling a non-vehicular form of movement, serving as makeshift stepping stones

Photo 9.3

Photo 9.4

that help residents to navigate the cityscape in areas vulnerable to flooding. Organized in linear fashion, the tyres have been repurposed in a manner that is both practical and temporary, symbolizing the transient nature of informal urban "fixes" in a rapidly changing environment. Yet the photograph prompts us to consider the fate of these tyres if the waters rise and overwhelm the narrow passageway—they may soon revert to their status as discarded waste, highlighting the fluid boundary between refuse and reuse.

We documented dozens of such informal uses across Lagos, some more durable than others—from goalposts for football games (Photos 9.4 and 9.5), to stands for the display of market goods, to weights helping to stabilize roofs (Photo 9.6), to informal means of marking out space in car parks. It seems likely that, in many cases, these functions benefit from the "anonymous" qualities of the biographies of tyres, which are readily abandoned without a sense of loss on the part of the owner, and in doing so may form part of an urban commons oriented around an ambiguously useful resource. Tyres, retaining their shape but changing their function, are also often deployed as platforms for portable diesel

Photo 9.5

Photo 9.6

Photo 9.7

generators (Photo 9.7). This latter juxtaposition is particularly suggestive, since it points to and augments Brian Larkins's characterization of generators in Nigeria as "ambient infrastructures" (2016), which extend across both urban and rural spaces—homes, bars, churches, markets—and not only form an essential part of the country's energy supply, but also play a key role in the ongoing experience of the environment. So, if everyday Nigerian existence is increasingly mediated by generators, the working of such apparatuses is augmented through a further mediating infrastructure in the shape of tyres, whose physical qualities allow them to absorb the vibrations of the machine placed on top of them. However, the juxtaposition of these two technologies of everyday life is telling. While generators are all too visible and noisy in performing their primary function, tyres under such circumstances are grounded, partially hidden, converted from their original use into a "fix" that began as innovation and has now become common practice.

Central to the composition of Photo 9.8 is a discarded tyre, now deliberately rendered hyper-visible as it is conspicuously placed to cover a manhole that lacks its original steel cover—likely removed or stolen. This

Photo 9.8

image highlights once more the improvisational strategies employed by Lagosians to navigate and mitigate infrastructural deficiencies. The act of filling the gap with a tyre, a piece of waste redefined as a protective barrier, sheds further light on the necessity of finding immediate solutions to the hazards that permeate urban space. In this case, the fact that the tyre is damaged actually adds to its functionality: it retains salient qualities of weight and a "coherent-enough" shape but is unusable for its original purpose and is therefore less likely to be removed by any passer-by. Such creativity is not only applied to cases of abandoned or disintegrating tyres. Our final image for this section illustrates how they may also be "upscaled" as objects—converted into works of art or craft, or even into museum pieces. In Photo 9.9, tyres become platforms for stools that are placed in neat and parallel lines along a path leading to a swimming pool at a café/ beach lounge, a place of relative luxury. Their material identity is both retained (it remains obvious that they are tyres) and customized (they are painted and paired with a cushion, made to fit into a leisure aesthetic).

Photo 9.9

So far, and using Andrew's photos as an ethnographic lens, we have been focusing on tyres as objects subject to numerous afterlives—abandoned, unmade, reframed, upscaled—and engaging with the city in ways that go beyond the "expected" friction of tyre on road. We have highlighted the physical qualities of tyres, and how such qualities can be deployed in adapting them to new uses. Generally speaking, with the exception of Photo 9.9, they have been depicted in assemblages with other objects but not in touch with the human body.

We now turn our attention to two new, but related themes. One refers again to recycling, but in a different way: to the processes of repair of tyres that restore them to their *original* use, ensuring that they circulate not through new systems of value but back on to the roads of Lagos. The second refers to the embodied labour that goes into such restoration—the forms of care that deal not only with the "rebellious" qualities of hard rubber but also with the sometimes dangerous characteristics of the air that is used to inflate the tyre. In other words, we turn to the work of vulcanizers and the "pneumaterality" they engage with.

"We Seal Air": The Restorative Labour of Roadside Vulcanizers

> We really want to know how to produce tyres. We are only working on fixing the problems with tyres. We only do repair of tyres. How do we know how to produce tyres? If we know how to produce tyres, it will give us an opportunity to have a good standing in the society. (Vulcanizer Ojo, 50, Lagos)

In their examination of Maine's second-hand economy, Isenhour and Berry (2020, 294) refer to the "long-standing cultural valorization of production-based…labor in capitalist logics" and go on to critique dominant economic frameworks that do not acknowledge the value of work required "to reclaim, repair, redistribute, and reuse" (ibid.). Isenhour and Berry's wry perspective on capitalism and what counts as productive labour acquires an intriguing dimension in relation to work on the Nigerian "pneuma-city." An industry oriented precisely around the mass production of tyres and involving local branches of the multinationals Dunlop and Michelin was developed in the country in the 1960s, soon after Independence, but these factories were shut down in the first decade of the new millennium. The closures are still a matter of regret for a number of our interlocutors. One tyre artist noted: "Yeah, but the problem is that do we have any tyre producing company in Nigeria? We don't have. Michelin which we used to have moved to Ghana but here is their market." Today, "Tokunbo and recycled tyres account for about 80% of the Nigerian total tyre market which are mainly imported from Asian countries and are of low quality."[9] The result, according to one vulcanizer, is that:

> Truly a new tyre is supposed to be better than a used tyre. But we have detected that some new tyres are not as good as some used tyres. Some used tyres are more powerful than some new tyres. For instance, some used tyres from Japan and England are more powerful than some new tyres from China.

What he is expressing here is not appreciation for a product that gains venerability with age, but rather the stern realities of a political economy of tyre distribution where lower quality goods permeate a market lacking its own manufacturing base, or indeed much purchasing power by global standards.

These developments highlight the importance of labour oriented around extracting maximum use-value from a resource that is both essential and unpredictable in its continued functioning. Given the state of the tyres but also the road, virtually all vehicles in Lagos require the services of a vulcanizer, sometimes on a regular basis. Characteristic functions of the role include fixing punctures, checking pressures, replacing worn tyres—not necessarily with new ones, but with relatively *less used* ones—and storing small numbers of tyres until they are collected or given away. Such "restorative labour" is not productive of a new commodity but reproductive of an old one. Nonetheless, as illustrated by the urban choreography with which we began this chapter, vulcanizers do more than simply repair tyres: they enable their collection by agents of the formal or informal economy (and upcyclers as well as "tyre artists"), while also being at the centre of a contested project of urban governance, where every few months they are required to clear the streets of abandoned tyres during "sanitation" days, thus directly addressing the practices of abandonment depicted in our previous section.

Overall, then, the labour of vulcanizers occupies numerous positions in relation to the generation of value from tyres: repairing, selling, storing, cleaning up, passing on. It locates these workers in crucial positions in relation both to formal governance and informal economies, yet—rather like the tyres upon which they work—vulcanizers also traverse multiple regimes of visibility and invisibility in the city. Situated on the roadside, vulcanizers face numerous vulnerabilities ranging from the physical hazards of repairing and/or inflating large tyres, which can burst and cause serious harm, to the challenge of working in polluted environments. The difficulties of manipulating tyre rubber are often cited as one of the reasons why vulcanizing is a predominantly male activity, though we did encounter female members of the profession, including at least one who had taken on male apprentices. Most operate without the protection of a shop, leaving them exposed to the elements, especially during the rainy season. Additionally, their (low) income is largely unpredictable, varying significantly from day to day:

> Every day isn't profitable; some days we don't earn much, and other days we earn enough. Let's say between 5,000 nairas [3.5 US dollars] and 10,000 nairas a week. (Vulcanizer Adam, 35)

Furthermore vulcanizers, in common with other urban actors including street traders and transport workers, navigate a challenging environment characterized by informal rent extraction from various sources—from local "area boys,"[10] the police and other agents operating on the public space. However, unlike the mobile transport workers studied by Daniel Agbiboa (2022), who face incessant predatory harassment from bus station/stop "tax" collectors as they move across Lagos, vulcanizers benefit from their relatively stationary position. Their local anchoring allows them the opportunity to develop localized approaches to negotiating and managing the everyday landscape of rent extraction. Self-governing strategies play a crucial role in this process. Through their association (the Lagos State Vulcanizers Association, LSVA), operating across well-defined zones, vulcanizers can obtain the support of their senior leaders who play a pivotal role in assisting them with the intricacies of local governance within urban informal spaces. For instance, these leaders facilitate the establishment and relocation of vulcanizers when required and mediate their interactions with garage owners, landlords, police and traffic officers or environmental agents. This mediation ensures a degree of protection and stability for vulcanizers, who operate in a precarious environment subject to frequent efforts by the state, urban elites and even private developers to "formalize" —that is, control, discipline or even suppress—informal economies and spaces in megacity Lagos (Fourchard 2021; Nwanna 2015; Olajide and Lawanson 2022).

While these associational dynamics rely heavily on social control through strict rules and regulations—for instance about pricing and internal competition—and are intertwined with patron-client relations (prevalent across most "informal" economic sectors),[11] they are a great source of pride for many vulcanizers we interviewed. These vulcanizers frequently speak of unity and collective support—"we are a family" (vulcanizer John, 45)—highlighting the sense of solidarity and mutual aid fostered by their associational networks even across religious and, to a lesser extent, ethnic boundaries (vulcanizers in Lagos are mostly Yoruba) (Photos 9.10 and 9.11).

In addition, the narratives we have collected emphasize the challenge vulcanizers face in redefining roadside work from degrading and precarious, to essential and skilled labour, mainly through a presentation of the crucial role they play in the city. As "doctors of tyres"—as some of them put it—vulcanizers patch the city, allowing it to function while recasting the tyre as central to the infrastructure of everyday mobility. One of the

Photo 9.10

Photo 9.11

mottos of LSVA states: "No tyre, no movement." This self-perception aligns with insights into the socio-material importance and history of rubber and tyres (Tully 2011). In her fascinating ethnography of "Michelinville" (Clermont-Ferrand, the city of the Michelin tyre company in France), Corinne Védrine (2019) shows to what extent the tyre came to be represented as an object of modernity, the key node of a wider, kinetic infrastructural network encompassing roads, signposts, maps, guides, garages, and repair stations. From an early stage, Michelin strategically encouraged the creation of this totalizing system[12] to ensure the widespread use of cars, facilitating not only the practical aspects of driving but also embedding the car within everyday life, thus promoting continual tyre usage and replacement. The "Michelin myth" she analyses, partly based on the paternalist and sacralized "father" figures of André, Édouard and François Michelin,[13] recasts the tyre company as a "demiurge," creator of a "true cosmogony" of a holistic, globally interconnected automobile world (2019, 47), which the tyre (and tyre-based innovations) has helped to propel, sustain and expand.[14]

Much has already been said in this chapter about the physical characteristics of the tyre. This "pneumateriality" is also crucial when considering the role of vulcanizers in relation to the matter of air. When asked about the major technical changes in tyre maintenance and repair they have experienced over the last few decades, most vulcanizers mentioned the shift to tubeless tyres (on cars in particular). Unlike their tube-type counterparts, tubeless tyres allow for a delayed deflation process, enabling them to remain functional longer despite punctures, including those induced by the nails that are common during the rainy season as debris washes up on Lagos' roads. This technology also has profound implications for the *materiality of air*, the relationship between vulcanizers and their craft, and their position in the kinetic world of the megacity. In a way, tubeless technology creates the possibility of a liminal state where tyres can be neither fully functional nor completely deflated, demanding regular but perhaps less urgent attention from vulcanizers. More generally, air as an active participant in the lifecycle of tyres necessitates constant monitoring and management by vulcanizers, who must then contain small leaks through their restorative labour—"we seal air" as vulcanizer John put it—to ensure continued mobility. The materiality of air in this context is characterized by the pressures it exerts within the confines of the tyre, as well as the dangers associated with exploding tyres.

Our small digression on the materiality of air points to its broader infrastructural significance. Historically, air has been associated with both life-sustaining and disease-carrying properties, as shown by Alain Corbin (1982/2016) classic *Le Miasme et La Jonquille*, examining the representations of air as vector of disease and social abjection (through smell and "miasma" for instance) and tracing the birth of sanitation through deodorization in eighteenth- and nineteenth-century urban France. Corbin's analysis reveals the dual agentic potentiality of air in the early industrial and modern urban world—both enabling and challenging what hygienists and planners saw as "human progress." In the context of tyres, it is not merely the *circulation* of air (seen as crucial by hygienists to maintain socio-spatial boundaries and domestic/bourgeois respectability) but its *containment* and *compression* that are pivotal. Vulcanizers regulate this "pneumateriality," ensuring the functionality and efficiency of urban transport systems and vehicular movements.

Our analysis concurs with Ingold's (2010, S132) argument that work on materiality has been confined to the "congealed forms of the landscape" and that "solid objects resting on its surface" leaving air literally *unthinkable* within these limits. Ingold insists on readmitting air as an essential material constituent of the inhabited world, as a medium that "affords locomotion, respiration, and perception" (ibid.). Thus, vulcanizers are not merely interacting with the (external) frictional materiality of tyres but are managing the dynamic flows of air within them. For vulcanizers, air is materiality because it is relational: contained by the rubber tyre but also acted upon by external forces, such as the weight of the vehicle or the road (here the rubber tyres mediate this relation), as well as the compressor, which alters the physical characteristics of air.[15] We need also to remember that the hazards associated with high-pressure air make the vulcanizers' work risky. The complex and often precarious nature of vulcanizers' engagement with air requires them to navigate both its affordances and its destructive potential.

Concluding Observations

Tyres take on many lives, shifting shape, value and location as they move into, across and sometimes out of a mega-city such as Lagos. We have traced some of the ways in which they are remade, reframed and repaired as they circulate—literally and metaphorically, formally and informally—through the multiple streets and varied economies of an urban realm that contributes to, and is formed by, global connectivities. In the process, we have observed how tyres may move between industrial processes entailed in mass production and destruction on the one hand, and more intimate, smaller-scale forms of craft that include both the care of the vulcanizer and the creativity of the artist.

In highlighting the multiple dimensions of tyres as made and unmade (as well as slowly decaying) objects, we have indicated the ethical and ecological dilemmas raised by their production and use. Clearly, there are deep environmental costs involved not only in the making of tyres but also in their contribution to multiple forms of pollution. Furthermore, the tyre can only work *as* a tyre by becoming worn down, gradually releasing invisible particles into an atmosphere that is itself both material and unseen. The resistant characteristics of rubber are likely to make an abandoned tyre a form of waste that remains "rebelliously" within a street or ditch before being "sanitized" by a vulcanizer; but such resilience may also contribute to the tyre's environmental redemption as it is converted into a new form of object. Indeed, the term "second-hand" may be too limiting for the purposes of this chapter, given that a tyre is likely to undergo third or fourth or even more "hands" as it traverses different spheres of utility.

Our use of the word "friction" in this chapter both echoes and departs from that made famous by Anna Tsing (2005). We do not focus especially on her themes of the awkward yet potentially productive qualities of interconnection between very different cultural actors. Nonetheless, we are, like Tsing, cognizant of the ways in which larger, seemingly abstract forces—neo-liberalism, globalization, scientific knowledge, industrial processes—collide with and are realized through "local," in our case deeply urban, contexts. We also show how tyres bring together the frictional and the infrastructural (as well as the frictional and the apparently fluid) in contexts where tyres perform work, enable work and are the objects of work. The approach we have taken is therefore less "cartographic" *per se*—abstract, seen from above—and more concerned with investigating a socio-spatial landscape on the ground: not one perceived from a single point of view but rather one viewed from numerous perspectives, revealing multiple forms of sociality and spatiality. The notion of landscape affords the possibility of heterogenous objects brought together by a unifying theme; it also emphasizes the significance of vision itself as an act of navigation, recognition and sometimes appropriation. In our analysis, through text and images, we have brought tyres to the fore of attention: not necessarily as objects of beauty (though they may achieve that status), but as material aspects of human lives, whether in Lagos or elsewhere, that deserve to be seen in all their forms.

Notes

1. In this sense, the treatment of such material illustrates Caroline Knowles's (2017, 291) distinction between focusing on objects per se or understanding them as a collection of materials. Under certain circumstances, recycling entails an "unmaking" of the object.
2. Itself a combination of oiled-derived components (carbon black), synthetic rubber, natural rubber and additives.
3. Funded in 2019 by the British Academy, www.pneuma-city.org.
4. Most of the research was carried out in Lagos State with additional fieldwork in Ibadan. The data was collected through participant and street observation, sensory ethnography (photography and audio-visual ethnography), interviews with roadside tyre repairers (60 interviews), recyclers/upcyclers (including waste tyres upcyclers and artists) (20 interviews), and additional interviews and workshops with various stakeholders of the solid waste chain in Lagos. The majority of interviewees cited in this chapter have been anonymized through the use of pseudonyms.
5. Vulcanization refers to the treatment of natural rubber with sulphur, improving rubber's resistant properties (the process has gradually evolved to include the treatment of synthetic rubbers with other curatives).
6. Esiebo conducted visual ethnography for the Pneuma-city project, http://www.andrewesiebo.com/.
7. Certain human agents within the waste tyre chain, in particular those dedicated to collecting, sorting and cutting the tyres in poorer parts of megacity Lagos, are often migrants from other parts of Nigeria or neighbouring countries such as Benin. The "informal" and precarious dimension of their job reflects the outwardly demeaning status of working with waste—which today, as in colonial times, remains an interpretative category mediating moral evaluations and reinforcing extraneity and Otherness (Newell 2020).
8. "419" refers to the section of the Nigerian Criminal Code dealing with fraud.
9. https://independent.ng/nigerias-5bn-tyre-production-industry-requires-investors/.
10. A term describing young men who engage in various forms of informal, often illicit, activities and who hang around in public spaces and street corners, sometimes simply performing odd jobs for compensation.
11. Mostly because trade associations can be used to mobilize their members during elections.
12. For instance through the funding of road infrastructures and signposts across France and the publication of maps and of a (now renowned) restaurant guide.
13. Édouard and André Michelin founded the company in the 1890s and Edouard's grandson François (1926–2015) led it from 1959.

14. A process which in turn, one could argue, has contributed to a "planetary urbanization" *à la* Lefebvre (2003).
15. For instance, compressing air makes its molecules move more rapidly, which increases its temperature.

References

Agbiboa, D. E. 2022. *They Eat Our Sweat: Transport Labor, Corruption, and Everyday Survival in Urban Nigeria*. Oxford: Oxford University Press.

Alexander, C., and J. Reno, eds. 2012. *Economies of Recycling: The Global Transformation of Materials, Values and Social Relations*. London: Zed Books.

Appelgren, S., and A. Bohlin. 2015. Growing in Motion. *Culture Unbound* 7:143–168.

Carrier, J. 1995. *Gifts and Commodities: Exchange and Western Capitalism Since 1700*. London: Routledge.

Choplin, A. 2019. Cementing Africa: Cement Flows and City-making Along the West African Corridor (Accra, Lomé, Cotonou, Lagos). *Urban Studies* 57:1977–1993.

Corbin, A. 1982/2016. *Le miasme et la jonquille: L'odorat et l'imaginaire social, XVIIIe-XIXe siècles*. Paris: Champs Histoire.

Douglas, M. 1966. *Purity and Danger: An Analysis of the Concepts of Pollution and Taboo*. London: Routledge & Kegan Paul.

Fourchard, L. 2021. *Classify, Exclude, Police: Urban Lives in South Africa and Nigeria*. Hoboken, NJ: John Wiley & Sons.

Fredericks, R. 2018. *Garbage Citizenship: Vital Infrastructures of Labor in Dakar, Senegal*. Durham: Duke University Press.

Gandy, M. 2024. *Natura Urbana: Ecological Constellations in Urban Space*. Cambridge, MA: The MIT Press.

Ingold, T. 2010. Footprints Through the Weather-World: Walking, Breathing, Knowing. *Journal of the Royal Anthropological Institute* 16:S121–S139.

Isenhour, C., and B. Berry. 2020. "Still Good Life": On the Value of Reuse and Distributive Labor in "Depleted" Rural Maine. *Economic Anthropology* 7:293–308.

Knowles, C. 2017. Untangling Translocal Urban Textures of Trash: Plastics and Plasticity in Addis Ababa. *Social Anthropology* 25 (3): 288–300.

Kopytoff, I. 1986. The Cultural Biography of Things: Commoditization as Process. In *The Social Life of Things: Commodities in Cultural Perspective*, ed. A. Appadurai. Cambridge: Cambridge University Press.

Larkin, B. 2016. Ambient Infrastructures: Generator Life in Nigeria. *Technosphere Magazine* November 15. https://technosphere-magazine.hkw.de/p/Ambient-Infrastructures-Generator-Life-in-NigeriafCgtKng7vpt7otmky9vnFw.

Lefebvre, H. 2003. *The Urban Revolution*. Minneapolis: University of Minnesota Press.

Levitt, P., and S. Merry. 2009. Vernacularization on the Ground: Local Uses of Global Women's Rights in Peru, China, India and the United States. *Global Networks* 9:441–461.

Micklethwaite, P. 2021. *Beyond Recycling*. London: Routledge.

Newell, S. 2020. *Histories of Dirt: Media and Urban Life in Colonial and Postcolonial Lagos*. Durham: Duke University Press.

Nwanna, C. 2015. Gentrification in Nigeria: The Case of Two Housing Estates in Lagos. In *Global Gentrifications: Uneven Development and Displacement*, ed. L. Lees, H. B. Shin, and E. López-Morales. Bristol: Policy Press.

Olajide, O., and T. Lawanson. 2022. Urban Paradox and the Rise of the Neoliberal City: Case Study of Lagos, Nigeria. *Urban Studies* 59 (9): 1763–1781.

Reno, J. 2009. Your Trash is Someone's Treasure: The Politics of Value at a Michigan Landfill. *Journal of Material Culture* 14 (1): 29–46.

Reno, J. 2015. Waste and Waste Management. *Annual Review of Anthropology* 44:557–572.

Salvaire, C. 2019. From Urban Congestion to Political Confinement: Collecting Waste, Channelling Politics in Lagos. In *Territory, Politics, Governance*. https://doi.org/10.1080/21622671.2018.1561324.

Smith, D. J. 2022. *Every Household Its Own Government: Improvised Infrastructure, Entrepreneurial Citizens, and the State in Nigeria*. Princeton: Princeton University Press.

Thurlow, C. 2022. Rubbish? Envisioning a Sociolinguistics of Waste. *Journal of Sociolinguistics* 26:386–403.

Tsing, A. 2005. *Friction: An Anthropology of Global Connection*. Princeton: Princeton University Press.

Tully, J. 2011. *The Devil's Milk: A Social History of Rubber*. New York: Monthly Review Press.

Védrine, C. 2019. *The Spirit of Capitalism According to the Michelin Company: Anthropology of an Industrial Myth*. New York: Palgrave Macmillan.

CHAPTER 10

The Immense Possibilities of Six Yards of Cloth: Saris, Recycling, and Power in Indian Households

Subhadra Mitra Channa

Introduction

In this chapter, I examine two connected themes: recycling and the domestic domain. Recycling can be understood in many ways, it helps conserve the environment and is an activity that extends the lifespan and life cycle of an object that passes through multiple incarnations (Kopnina 2011). Recycling is a way to expand a resource, its capabilities, and the power that its possession bestows upon its owner. In the sense of conservation, one can refer to the present trend in western countries of wearing second-hand or recycled clothes, as a means to cut down on fabric consumption, now realized to be detrimental to the environment. What is a contemporary trend in the wealthy western countries has always been a necessity for large numbers of poor in the less advantaged countries of the Global South, where even today many people cannot afford new clothes or other consumer goods. A new dress or pair of shoes is a luxury for

S. M. Channa (✉)
Delhi, India

P. Kuppinger (ed.), *Second-Hand Cultures and Economies of Reuse, Repair, Sharing, and Care*,
https://doi.org/10.1007/978-3-031-99875-1_10

millions across the globe. For those who can afford to buy new clothes, wearing used clothes is a statement and has an environmental goal.

Recycling

Recycling for environmental concerns is increasingly visible in the context of art and manufacturing, where today many well-known fashion brands proudly advertise that some of their clothes are made from "recycled" materials. People are also encouraged with small incentives, to leave their used clothes at the counters of retailers such as H&M who promise to recycle them, no doubt making a good profit for the manufacturers who receive the materials for a pittance. One needs to distinguish between recycling that is done as a necessity and recycling done as a statement, although perhaps the result is similar, that is, conserving materials or diverting them from landfills. At the India Art Fair in New Delhi in February 2024, an art installation made of used cardboard packaging was announced to be donated to a school to insulate their roof. The artist said that there was a double transformation of materials: from packaging to art to insulation. In the Indian city of Chandigarh, the artist Nek Chand created a picturesque rock garden totally made from waste materials. These are conscious and deliberate methods of recycling that understand and use the concept of recycling purposefully and offer a contrast to recycling done by those who are forced by poverty.

Recycling as the expansion of a resource beyond the use for which it was initially created may happen either with political or economic motives, which can be determined through the methodology of recording the life history of an object. Consumption studies that examine the environmental impact of products have focused on the reuse of materials (Burn 1991; Heckler 1994; Oskamp 2000). Some studies are based on psychological motivations or the need to foster changes in attitudes that will then lead to changes in lifestyle. This chapter explores values, motivations, and especially aspects of satisfaction and fulfilment of higher order needs like aesthetic values, companionship, and affection that stimulate practices that result in the conservation of materials but whose goals were not primarily perceived as such. The conservation and recycling carried out by women in India is a part of life worlds where the reality of limited possessions makes each of them precious, to be utilized to the maximum and treated with utmost respect. This applies to scarce food or unavailable and costly items like books and textiles. Such values are learned early in life through religious and social value systems. As a child, my mother told me that each grain of rice was precious because it was produced by the hard labor of the peasant. Rice symbolizes the goddess Lakshmi[1] and should not be touched by feet or otherwise disrespected. Similarly, books symbolized the goddess Saraswati[2] and therefore should be treated with respect, should not be thrown around, torn, or kicked. Scarce resources are elevated to the status of the sacred to be revered and conserved.

If something can be made more useful, its usefulness extended beyond its initial property or functionality, then the power of this object as a possession increases. The term "resource magnification" best describes this process. Here, the concepts of consumption, use value, and exchange value are crucial. If an object has the potential to be exchanged it becomes a commodity. Appadurai (1986) gives a complex historical discourse on commodities drawing on Marx and other thinkers. He emphasizes social values, where the subjective and the objective intersect. The economic, social, and emotional value of an object is a compound of its physical attributes, in terms of material and design, and the emotional value of its source, like a gift and its history. Who gave it and under what circumstances? What is the relationship to the giver? What is the status of the giver? Are they dead or alive? The value of an object reflects the emotional weight of the relationship embodied in it. Was it a gift from a loved one, or from someone who is no longer alive (so it has additional value of memory)? Is it a rare commodity (economic value)? Does it have a history (social value) and artistic and aesthetic value (depending on the desire it evokes in others)?

Recycling means that the same object moves between different regimes of values, which can be use value or exchange value in barter or monetized exchanges. It can become a commodity or remain in the non-monetized social/cultural sphere. Tracing the life cycle of an object can provide important insights and indicates various aspects of power in different spheres of production, possession, and exchange (Kopytoff 1986). In this chapter, I focus on the sari, which is historically and socially constructed in terms of its transition from not just use to exchange value (which may not always happen), but transformations in use value and associated aspects, mainly social and psychological, where value becomes an inherent property of the object through the labor and emotion invested in it.

The Domestic Domain

The domestic domain has long been recognized as one without power or less power than the public domain because of its location within the women's sphere as opposed to the male realm of the public (Ortner 1974).[3] What is domestic and what is public is not a clear dichotomy in both non-capitalist and capitalist economies (Channa 2015). In South Asia and other parts of the world, goods and services may be produced and provided by households rather than by individuals. Items that circulate and enter the market may have been produced in the domestic domain where women play key roles in their production. For example, women make clothes and objects embellished with gold thread embroidery in their homes but these items are traded in markets by the men of the household (Mondal 2013). Some Muslim women do not physically enter the public domain because of social restrictions yet they participate indirectly in the market through their products.

The domestic domain can be a seat of power for women who, while being confined to it, expand its scope by indirectly participating in the public realm, like male-dominated markets. A woman may go a long distance to fetch water or get food and fodder from the forest, yet she will be recognized socially as doing domestic work. Similarly, I described the work of women of the washerman caste (Dhobi) that involves much movement away from their homes and their presence in public spaces and the handling and keeping of money, yet it is classified as *ghar ka kaam* (work within the domestic domain; Channa 2013a, 123).

The domestic realm can generate economic and political power. Some women produce objects of considerable commercial value in their households, like gold embroidered cloth or carpets woven by Bhotiya women on the higher reaches of the Himalaya (Channa 2013b). The domestic can further empower women if they are able to enhance the value of any object they exclusively own to use or dispose of. Radcliffe-Brown (1950) called these *rights in rem*. The owner can maximize her power by stretching the value of her possession by investing her imagination and labor into it. Women with the least possessions might use their wisdom, skills, and imagination and draw upon their feminine networks, to acquire power even within a patriarchal context that otherwise stifles their attempts at liberation, agency, and sense of self.

In this sense, the "domestic domain" is a concept metaphor as understood by Moore (2014, 364) which is a shared term, mainly among anthropologists, but whose exact meanings can never be specified unless they are contextualized, their empty spaces filled with details of observations, facts, and comparisons. This keeps open possibilities of debate and discourse and future ways of understanding and analysis. Gender specialists and ethnographers have filled the domain of the domestic, with a variety of meanings, taking it far beyond the initial Eurocentric meaning where it meant that this domain conferred no power and acted as a restriction for those who occupied it, namely, women. Such assumptions have been criticized using ethnographic examples that show that women wield considerable power both within the domestic domain as well as exert influence over the public domain.

Ordinary consumer articles and women's products made in the household can generate income and provide emotional and aesthetic fulfilment. Women, having limited access to resources, because of their limited participation in the public domain, create ways to stretch their resources through conservation and recycling. By doing this, they fulfil traditional expectations placed on them as custodians of their household's welfare. In India, women who marry into a household (daughters-in-law) are referred to as Lakshmi, but only when they engage in appropriate behavior that conserves and enhances household resources. A woman in charge of the household is responsible for managing resources strategically and maximizing them. A woman, who does not waste, makes maximum usage of every drop, and is not wanton in her behavior is regarded as the ideal *bahu* (daughter-in-law) and can become a powerful female head of the household. In India, the domestic sphere is headed by women and men are kept excluded: "Feminism in India must account for the fact that instead of women being directly under the control of men, here they are under the domination of older women" (Channa 2013a, 45).[4]

Women's practices of conservation and recycling are therefore empowering and translate into highly prized virtues. In situations of scarcity, this virtue requires that a woman minimizes her own needs and reduces her own consumption. A woman may starve herself or deprive herself of essential nutrition to feed, especially male, household members. Social power can come with physical disabilities from which older Indian women often suffer. As I have indicated elsewhere (Channa 2013a, 94) women in India are entitled to a high degree of prestige and respect and wield considerable moral power over men, but at a high personal cost. For clothes, this means having only the minimum and using items until they are frayed and worn out.

The Sari

The sari is one of a woman's most highly valued possession. By maximizing the usefulness of these "six yards" of cloth, women, even if poor and confined to domestic life, have managed to extract some power, in material, expressive, and aesthetic terms. The material nature of the sari can be reconstructed into a variety of objects, and its usefulness extended beyond its original use. This can add considerable value to a sari, which translates into power gained by a woman, who is able to enhance and enrich this piece of cloth through her imagination and industry.

A sari is a general category of dress with an expanse of five and half meters or six yards, which is draped in particular ways by South Asian women. A traditional sari has three parts, the base, the border, and the *pallu* or *aanchal*. The base is the color and design of the length and breadth of the material, which usually has a border on both sides that forms the edges of the sari. The *pallu* is the last part, about a meter in length that hangs prominently at the back and can be pulled over the head. The "look" of a sari is a combination of the three elements. Borders and *pallu*s are the most creative aspects of saris that represent the skill and imagination of the weaver. Different saris are marked by the way their borders are made. Border and *pallus* have different, often splendid, designs and names.

Different saris take their names and identities from borders and *pallus*, for example, the Temple border of South Indian saris. There are regional trademarks of saris, as parts of South Asia produce their own "brands" of saris. They often carry regional names and are characterized by special kinds of materials, designs, crafts, and technologies. For example, in the pre-colonial era, one of the most famous saris, the Dhakai Muslin, was made in what is today Bangladesh. This is a cotton sari so gossamer that it was famed to be able to pass through a finger ring. This sari was woven in magnificent designs known as *jamdani*. This material is no longer widely produced. If manufactured, it is very expensive and only for elite wear. The weave remains prevalent, has high aesthetic value, and is only woven in superior materials like fine cotton and silk for formal wear. Dhakai saris are still made by traditional weavers in Bangladesh and traded across the world at high prices, but they no longer pass through finger rings. In neighboring West Bengal, India, the technology of *jamdani* is used by weavers to make saris that are similar to but not identical to the Dhakai. These saris are also expensive and meant for ceremonial and fashionable wear. Other famous sari making centers are in Orissa using unique designs and color combinations in designs known as Ikat, Sambalpuri, or Bomkai. Benaras has rich weaves of gold embroidered silks; Kanchipuram in South India is famous for its distinctive heavy silk saris known to look new for generations. Maharashtra and Gujarat also have their own distinctive saris that can easily be distinguished. They are made by traditional weavers who have perfected the art of weaving over generations. Most women can easily distinguish these patterns and regional origins.

These traditional woven saris are becoming very expensive because of the cost of organic materials like fine cotton and silks. Today, machine woven saris using a variety of cheaper synthetic materials are more affordable for economically marginalized women. These saris are referred to as mill made and are not specifically associated with any design or color combination like the traditional weaves of the handloom-made saris. Some mills also make cotton and silk saris and have developed their trademark designs like one mill making lacy cotton saris and another making specific soft cotton saris for home wear. These saris have acquired names and are recognized as suitable for specific occasions and purposes.

Gifting, Using, and Reusing Saris

Although the sari may appear to be always the same dress, it assumes the role of many dresses, and some saris are marked for special purposes and occasions. Some are worn only for ritual occasions and ceremonies. Every region in India has a special bridal sari, unlike those of Christian brides that vary only according to personal taste. A bride in Kerala wears a white cotton sari with gold weave, a bride in Bengal wears a red and gold woven sari from Benaras, and a bride in Rajasthan wears a Gharchola sari, which has a regional flavor.

Women who wear saris at home prefer coarse cotton or synthetic ones for easy maintenance. Cotton saris usually need to be starched and the washermen (Dhobis) are experts in making and putting starch along with the women who wear them. Saris are not draped in the same manner across India; every region has its own distinctive way of draping and sometimes the length and breadth of the sari varies according to the way it is draped. The "modern" sari that is often worn by women, whose images are cast in public, was invented in the colonial period. It was specially designed by women such as Sarojini Naidu and Kamala Nehru to enable women to participate in the independence movement and come out on the streets. This sari is worn with a petticoat and blouse, both English terms referring to articles of clothing borrowed from the British. It is draped in a manner that gives a "respectable" look and enables movement in public spaces (Channa 2013a, 57–58). Traditional forms of saris allowed for freer movement but did not conform to standards of "decency" required during the Victorian era of British rule. In Bengal, for example, the sari was worn without any inner wear, no blouse or petticoat, and rural women draped it in a way that left most of their arms and legs bare. With this drape, they were able to climb trees, swim, and even ride horses. Elite women never encountered unrelated males and covered their heads and shoulders, otherwise left bare, with the ends of their sari when they came across any elderly family members like their father-in-law. The modern sari is a symbol of elite women's emancipation from the confines of the domestic domain and entry to the public domain. Lower caste/class and rural women were always in public and had their own ways of saving their honor, not by physical modesty but strategies of manipulations. Many rural women continue to wear their sari as it was worn in the pre-colonial era. However, trends are changing fast, mostly due to the influence of the media.

Most women receive a new sari on ritual occasions or are expected to receive one. At their wedding, women expect to receive the largest collection of their lifetime. Even the poorest families try to give their daughters at least five new saris. For the case of the dhobis I studied (Channa 1985, 2024), these were often the collective gifts of the entire community. A wealthy bride may get a hundred saris. The value of these saris varies greatly across economic classes. Poor women are happy to get a sari worth fifty rupees, while the glamorous wife of India's billionaire tycoon Mukesh Ambani, Nita Ambani, reportedly wears saris worth millions of rupees each time she steps out. An average middle-class woman has at least ten ritual occasions every year when she can expect gifts of saris from her husband, father, brother, mother-in-law, and others.

Traditionally, women never bought their own clothes, they received them as gifts and it was only with the entry of women into the public sphere that they became empowered to buy their own clothes. However, what is referred to as a gift is not strictly one in the way described by Mauss (1976). They are more obligatory presentations and follow the time sequences as described by Bourdieu (1977). People keep a record of all gifts that were received at a wedding to be returned at the appropriate time and occasion. There are occasions when a woman receives a sari from a designated relative almost by right, although if the family is very poor, they may not be able to fulfil this obligation. There are also situations of pure gift giving, where a sari is actually gifted, but this is likely only among those who can afford to do so. In other words, a woman confined to the domestic sphere depends on gifts to have saris, although elite women did buy saris of their own choice that are paid by the men of the family. In the past, these were bought from weavers who came to the house to sell them. A woman who goes to the market to buy herself a sari is a recent phenomenon and not all women can do this. While the sari is seen as synonymous with womanhood in India (Bangladesh, Nepal, and Sri Lanka) it is not worn by women in some regions, namely, the North-West, North-East, and the desert regions of Rajasthan and Gujarat, which have their own distinctive dresses.

A Culture of Conservation

Even in the past when all inheritance was in the paternal line, a woman always had the right to ownership. They could own saris, jewels, and gold ornaments and could freely use and dispose of them. These personal possessions, especially those a woman received at her wedding are known as *stree-dhan* (woman's wealth) and she had complete control over them. Most Indian women do not hold paid jobs. Many have almost no education and yet are responsible for their family's welfare, mostly participating in the informal economy. For these women, these possessions are the only insurance that they have in moments of crisis. Women learn how to make the best use of these precious items and they make do with whatever resources they have and stretch them as far as possible. Since women have limited access to resources and even if they are earning modest amounts of money, they remain the primary caregivers for their families. While men may be irresponsible, women are very rarely so. They were socialized to be caregivers as exemplified to them by the older women in the family, and stories and mythology that inform Indian society.

Growing up and even now when I have been running my own household for more than five decades, I have never bought dishcloth, dusters, or paper towels. These items have always been made by recycling old cloth, often an old cotton sari, drape (*dupatta*), or *kurta* (top wear). I was shocked when I first arrived in the United States about the huge amount of paper being wasted (in my perception) on wiping and cleaning. Yet, this is only one, and a lower order of use and reuse. Most women in India have many more creative uses for their old clothes, especially saris. While cotton and cheaper saris are used for menial tasks like scrubbing and cleaning, expensive saris are recycled into fancier items. Soft cotton saris have provided mothers with resources necessary for babies, like for making cloth diapers (now replaced in the middle classes by single-use commercial diapers). The sari was torn up and used as wipes (now often replaced by commercial wipes), folded and used as spreads. A sari can also be converted into decorative items. A length of soft cotton material is a handy resource and can be reused in many ways. An old, yet still beautiful sari can be converted into a little girl's dress (I have done that for my daughters), into curtains, and bed covers. Stitched clothes like *dupattas* and *kurtas* (very popular among younger women) are now often remade into trendy dress items for the younger crowd. Mothers may satisfy the fashion craze of a teenage daughter by handing her one of her expensive saris to be refashioned into a dress. In a low-income household, this is a way for a young girl to fulfil her dream of being fashionably dressed, and she remains grateful to her mother for her sacrifice of maybe a favorite sari.

Trading, Downcycling, and Upcycling Saris

Traditionally, expensive saris included actual gold and silver threads in their embroidery and could easily be sold. Women had barter and exchange systems where lower caste women went from household to household, carrying kitchen vessels of stainless steel and plastic and exchanging these for old clothes, especially saris. Saris had the highest exchange value in this barter trade because of the vast range of possible uses. Saris, especially expensive ones with gold embroidery, have been highly priced. These women traders belong to a community known as Waghiri (Bapat 2024). They are called *bartanwalis* (vessel women) or Chindiwallis and have a penchant for gossip and much bargaining power. Every time a deal was made, it was a power game between the two women: the owner of the clothes and the *bartanwali*. I have participated in such exchanges, and they can take an hour or two of intense negotiations and both sides try to get the best possible rate for their ware. Ultimately, this barter exchange was much like a market exchange just without the use of money. Each side was trading an object that had a market or monetary value and while haggling, the women have this value in mind. In their negotiation, they emphasized the purported monetary value of their object. Some such barter relationship were long term ones, as a particular *bartanwali* became attached to a particular housewife and, although the bargaining was just as stiff, the quality of the good she offered improved. The relationship may even develop into a friendship where the traveling trader, who was walking long hours with her wares, would find a place to rest, get refreshed, and chat. Such barter relations between women were never a purely formal market exchange but always included the female traits of sorority and empathy.

With their social and sentimental dimensions, such barter was an efficient mode of recycling where women gave away old clothes and received something useful in return. Bapat provides a detailed description of how old clothes are upcycled and made marketable with a much enhanced exchange value in the course of such exchanges (Bapat 2024, 116). Old clothes were used by the trader or they entered larger exchange networks. Some pieces ended up in waste cloth circuits, where they became part of the capitalist market network if they were recycled into marketable products (Norris 2010). Some women acquire saris in this exchange system to recycle them into products such as cloth bags, scarves, items of clothing, and other products where the original design and decorative aspects are reinvented and utilized. These recycled products enter market networks where they find new uses and values. The original owner in this chain of exchanges has to be satisfied with whatever she was able to bargain for. However, later in the chain the value of her original sari may increase considerably in value depending upon the skill and artistry of the person who reinvents and recycles this piece of cloth. Much of this creative reworking and recycling is done by women, men may only enter at the exit point to the market.

Saris are often exchanged in an unbalanced reciprocity as part of complex exchange systems expressive of kinship and other forms of affective social bonds (Sahlins 1972). They have the unique characteristic that quite literally one size fits all. Thus, it is an item of clothing that can be easily exchanged and preserved over generations to be handed down. Such generational wear is yet another form of recycling. It is an emotional moment for a sari to be given to a loved one to be worn on a special occasion. It is an act of emotional investment and a symbolic gesture. For example, the wedding sari worn by India's late Prime Minister Indira Gandhi was woven for her by Gandhi and his wife Kasturba on the spinning wheel. This sari was later worn by her daughter-in-law, Sonia Ghandi, and her granddaughter for their respective weddings. Young brides recycle a loving grandmother's sari for their wedding as an emotional statement. Sonia

Gandhi, the Indian political leader, reportedly wears Indira Gandhi's saris, rather than buying her own. Like items of Kula exchange, saris carry their history and act as personal memory banks where their value is judged in subjective and not objective terms. In one house, I found a beautiful wall hanging and was told that it was the *pallu* of an old sari the woman of the house had artistically framed.

Recycling and Empowerment

Saris fulfil many important tasks and can be a source of empowerment. For example, in poor village households in Bengal, women layer worn out saris, that might be torn in places, and then stitch them together, keeping the fragile layers in place by close embroidery. Many women cannot afford to buy the thread for the stitching and embroidery. They recycle threads by pulling them out of the woven borders of the saris. I remember my mother doing this, not because of poverty, but it was a way of life not to waste anything or to throw away anything that could potentially have any use. To this day, I find it difficult to throw away anything that still looks that it could have some use. The finished product in the shape of an (multi-layered) embroidered throw, spread, or mat is known as *kantha* and often used to wrap babies or as a shawl.

The making of *kantha* is a leisure or social activity where women gather, exchange threads, ideas for designs, and share their grief, joy, and gossip. It is a woman's world where they are independent. The materials they use belong to them, as a sari is a woman's sole property and she can do with it as she pleases. She can wear, exchange, gift, recycle, or source it for other purposes. It is her decision what she does and her own creativity what she makes of it. The re–/use of saris allow women to express their emotions and aspirations and satisfy their aesthetic needs. Motifs stitched onto the *kantha* can express myths, stories, and narratives that are close to a woman's heart. Some are religious or tell social or emotional tales, like of a mother and child or a sister's longing to meet her brother. The *kantha* is only one example; all over India, women make beautiful things using recycled material, mostly their old clothes. They use the cloth in patchwork, layered, and embroidered, and creatively reconstructed into useful objects and impressive works of art.

More recently, many of these crafts and items produced by women entered the commercial market. *Kantha* is no longer something that is merely recycled, but the stitch and design have been incorporated into the production of fashionable and expensive saris and dresses, using expensive new materials like silks and fine cottons. A new genre of clothing has emerged known as *kantha*. Such stitch clothes and a beautiful sari using this embroidery technique can be pricey. These textile developments have empowered women who can now use their skills to create a cash income. They continue to use their old work groups of friends and family to do this time intensive work. As a group activity, this labor becomes pleasant, brings women together, and provides them with an income. Much of this income goes into the family funds as the saris and clothes are marketed by middlemen, who often also provide the raw materials. Or they are traded by male family members who treat the revenue as household income. Only few women who can control the entire process, from buying or sourcing materials to selling their own products, manage their own profits. Some women's self-help groups market their own products. Some talented workers join high-end fashion designers who are now promoting and using such traditional skills in the international fashion arena. Similar transformations, from household recycling activity to global markets, are underway with other skills elsewhere in India. Women add value to old clothes by embroidery and stitches, such as *phulkari* embroidery of Punjab or the Kutch technique of using glass and beads carried out by indigenous Kutch women in Gujarat and Rajasthan.

Recycling and Domestic Power

Historically women in India have often been muted. The higher the social status of a household, the more invisible women were. Less privileged women were often engaged in caste-based occupations that left little time for leisure activities or creative work (Channa 2001). Some, however, expressed their skills and creativity by recycling their possessions, especially saris. A sari's lifespan can be expanded and its value increased through processes of conversion and exchange. This work, often performed in groups, encourages sorority and bonding among women. Clearly, recycling is meaningful beyond its material and environmental significance.

Arjun Appadurai (1986) notes that a "thing" is not just material but has a social existence. Recycling is a way to change a "thing"; it can extend its lifespan, give it new meaning, and can confer economic, social, and emotional power to its owner. Recycling is "life giving" work that preserves the object, while magnifying its symbolic and material existence. In feminist terms, recycling is empowering and enriching. Ownership gives women power, and they can further this power by extending the lifespan of their possessions. Women can use their skills to enhance the value of a gift, like embellishing a plain sari with embroidery, painting, and making it something special or unique. This empowers the giver and marks her as an exceptional giver.

Moore notes, "cultural production and innovation are at the basis of the expansion of the economic system, and suddenly what is for sale is the production of everyday life" (Jameson 1998, 67 quoted in Moore 2014, 368). The corrosive impact of capitalism and its persistence in imposing a monetary economy transforms established practices done for pleasure and joy, into commercial activities. If a woman makes a *kantha* out of her own volition, for her own use and pleasure, she is getting an elevated experience. However, if she does the same work in the context of market production, she labors under structural limitations of time and exchange values and her work ceases to have the qualities of creativity and liberation and becomes a routine income generating activity. The pervasiveness of patriarchal values in the market robs her of the power she would have had if the activity were one, which she could control rather than being controlled by the forces of market-based exchange. In this sense what is promoted as elevating a woman's economic status and empowering her by bringing her creative skills to the market may really disempower her by inserting her work into global capitalist circuits. If new cloth is substituted for the recycling of old clothes, the associated values of conservation are also lost.

Concluding Remarks

In this chapter, I examined recycling as a form of production or reworking that involves the use of already existing materials. I described recycling in terms of holding on to an item to extend its usefulness, before it goes into a landfill or becomes garbage. Environmentally the world is sinking under the weight of enormous quantities of waste that end up in landfills or even the sea. Much of this waste is the result of overconsumption encouraged by global capitalism and the amalgamation of the economic and the social (Gudeman 2009) where one's social identity becomes inextricably linked to one's economic wealth and status. However, in India, such social prioritization of the economic as a key to status identity is restricted to a largely westernized elite. Yet, there are plentiful historically inherited cultural practices that delay the process of "turning into waste" for a prolonged period. These long-established recycling practices are social and cultural but today also constitute relevant aspects of environmental work and efforts. In India and elsewhere, women, even in the middle classes, in their efforts at good housekeeping, often play key roles in recycling. While for more affluent women, it is a matter of pride to be an efficient housewife, for the poor, it is a matter of survival for their families, and of their personal dignity and self-care. Recycling is a crucial element in their management of scarce resources.

In the case of Indian women's textile recycling work, it is important to understand that these practices are rooted in longstanding cultural values of respect for goods and materials, often embedded in ritual language and the resulting concrete work of treating these materials with great care and making them last as long as possible. Reuse and recycling continue to be crucial parts of many women's daily work but often done as routine practice rather than with any specific thought or deliberation. I illustrated how saris are reused when they are gifted, especially across generations and showed how woman make children's clothes or fashionable outfits for their older daughters. Possibilities are endless, like saris being converted to bed cover, shawls, purses, and other items. Furthermore, I pointed out household practices of recycling and downcycling whereby women reuse old materials as diapers, wipes, rags, or related uses. Very little of these practices are rooted in contemporary ideas of recycling; they vastly precede them. In the end, however, recycling practices—regardless of their roots or motivations—all help to keep materials from landfills and as such have ecological results.

Using the examples of *kantha*, I explored how time-honored practices, often rooted in contexts of poverty and resulting frugality have in recent decades been discovered by (global) fashion industries that are now eager to employ women for these labor-intensive tasks. Elements of what once were admirable skills that poor women had to master have been absorbed into global capitalist circuits and made their entry into other markets. In the process, women who did the work for aesthetic pleasure or as a pastime, using material to which they had easy access, found that this skill could be marketed, thereby giving them some agency and income. However, at a higher level of production and marketing, women often have to take the back seat, letting men do the front desk job. Increasingly, their labor is hijacked by name brand designers. Yet even in such situations, they can gain economically and socially. Some women have become entrepreneurs themselves and they provide economic opportunities to other women by way of ***kantha*** work. *Kantha* stitch's conversion to a fashion statement and its transition from a humble household practices and good to an expensive marketable commodity has contributed to a limited extent to women's empowerment and agency, and their families' welfare. However, the large-scale merchants who sell women's products in global markets make most of the profit for these elaborate products.

While recycling in its different social, economic, and historical contexts is a positive activity that makes the best or long-term use of materials, observers should not lose sight of contexts where recycling work, performed by disadvantaged groups, is used and claimed by others. The global market for second-hand clothes/saris, which begins with the door-to-door barter of marginalized communities, involves women who derive little economic advantage from their work. Such processes might be ecologically beneficial but to be socially sustainable they need to pay attention to the process of redistribution and profit sharing that remains heavily biased in favor of those at the top of the chain. At the same time, we should not lose sight of the vast ecological contribution of relatively invisible groups like ordinary women in India, who engage in much recycling work without situating it in discourses of the environment and sustainability. Their contributions are tremendously valuable and must be given more recognition than they are receiving at present. The state somewhat helps here with the organization of direct markets involving the craftsperson especially women, where they can exchange their products directly without involving agents or merchants, but these are still not enough. The women and these markets are beyond the gaze of the international community, including environmentalists. Overall, the tremendous ecological contributions of especially marginalized Indian women remain largely invisible and unappreciated.

Notes

1. Lakshmi is the goddess of wealth and prosperity of the Hindu Pantheon, where wealth implies good fortune and wellbeing. The daughter-in-law of a household is also referred to as Lakshmi as she symbolizes the good fortunes of the household.
2. Saraswati is the goddess of learning, knowledge, arts, and skills.
3. Although a lot has been written on the public/private dichotomy since then, Ortner's was the seminal article that showed that women's reproductive and nurturing roles were assigned less value than men's activities outside

of the domestic sphere. What men do confers more prestige and power, following the Renaissance philosophy, Nature (feminine) as inferior to Culture (masculine). This is a Eurocentric view and does not apply to many non-western societies (Channa 2015).

4. For more details about this separation of the world of men and women, see Channa (2013a, 42–50).

References

Appadurai, Arjun. 1986. Introduction: Commodities and the Politics of Value. In *The Social Life of Things*, ed. Arjun Appadurai, 3–63. Cambridge: Cambridge University Press.

Bapat, Dipti. 2024. Moving With Rags: India's Second-hand Clothes Recycling Trade. *South Asian History and Culture* 15 (1): 111–129.

Bourdieu, Pierre. 1977. *Outline of a Theory of Practice.* Cambridge: Cambridge University Press.

Burn, S. M. 1991. Social Psychology and the Stimulation of Recycling Behavior: The Block Leader Approach. *Journal of Applied Social Psychology* 21:611–629.

Channa, S. M. 1985. *Tradition and Rationality in Economic Behaviour.* Delhi: Cosmo Publications.

Channa, S. M. 2001. The Right to Selfhood: The Paradox of Being a Dalit Woman. *Social Action* 51 (4): 337–351.

Channa. 2013a. *Gender in South Asia: Social Imagination and Constructed Realties.* Cambridge: Cambridge University Press.

Channa. 2013b. *The Inner and Outer Selves: Cosmology, Gender and Ecology on the Himalayan Borders.* Delhi: Oxford University Press.

Channa, Subhadra Mitra. 2015. Men as Nature and Women as Culture: Women's Work in a Pastoral Community on the Himalayan Borders. In *Gender, Livelihood and Environment: How Women Manage Resources*, ed. Subhadra Mitra Channa and Marilyn Porter, 27–49. Hyderabad: Orient Black Swan.

Channa, Subhadra Mitra. 2024. *The Dhobis of Delhi: An Ethnography from the Margins, 1974-2023.* Oxford University Press.

Gudeman, Stephen. 2009. Necessity or Contingency: Mutuality and Market. In *Market and Society: The Great Transformation Today*, ed. C. Hann and C. Hart. Cambridge: Cambridge University Press.

Heckler, S. E. 1994. The Role of Memory in Understanding and Encouraging Recycling Behaviour. *Psychology and Marketing* 11:375–392.

Jameson, Frederic. 1998. Notes on Globalization as a Philosophical Issue. In *The Culture of Globalization, Frederic Jameson and Masao Miyoshi.* Durham, NC: Duke University Press.

Kopnina, Helen. 2011. What About That Wrapper? Using Consumer Diaries in Green Education. In *Environmental Anthropology Today*, ed. Helen Kopnina and Eleanor Shoreman-Quimet, 118–139. London and New York: Routledge.

Kopytoff, Igor. 1986. The Cultural Biography of Things: Commoditization as a Process. In *The Social Life of Things: Commodities in Cultural Perspective*, ed. Arjun Appadurai, 64–91. Cambridge: Cambridge University Press.

Mauss, Marcel. 1976. *The Gift*. New York: Norton.

Mondal, Sekh Rahim. 2013. Gendering in Traditional Industries: Some Observations on Hand Embroidery (*Jari*) and Coir (*Bolen*) Works of Bengal. In *Gendering Material Culture: Representation and Practice*, ed. Subhadra Mitra Channa and Kamal K. Misra, 115–128. Jaipur: Rawat Publications in Association With the Indira Gandhi Rashtriya Manav Sangrahalaya.

Moore, Henrietta. 2014. Global Anxieties: Concept-Metaphors and Pre-theoretical Commitments in Anthropology. In *Anthropology in Theory*, ed. Henrietta L. Moore and Todd Sanders, 363–376. Wiley Blackwell.

Norris, Lucy. 2010. *Recycling Indian Clothing: Global Contexts of Reuse and Value*. Indiana: Indiana University Press.

Ortner, Sherry. 1974. Is Female to Male as Nature Is to Culture? In *Women, Culture and Society*, ed. Michelle Z. Rosaldo and Louise Lamphere. Stanford: Stanford University Press.

Oskamp, S. 2000. Psychological Contributions to Achieving an Ecologically Sustainable Future for Humanity. *Journal of Social Issues* 56:373–390.

Radcliffe-Brown, A. R. 1950. Introduction. In *African Systems of Kinship and Marriage*, ed. A. R. Radcliffe-Brown and Daryll Forde. London: Routledge and Kegan Paul.

Sahlins, Marshall. 1972. *Stone Age Economics*. London: Tavistock.

CHAPTER 11

Reclaimed Wood in a Reclaimed City? Or, the Story of Joe's New Table

Simon Johansson

During the last decade, the circular economy (CE) has emerged as an important imagination in environmental, economic, and urban policy worlds. As it gained wider traction in academia, industries, and among policy professionals, its conceptual boundaries became blurred, especially with regards to the concept of sustainability (Geissdoerfer et al. 2017). Although there are multiple ways to define and manifest circular economies, many rest on the notion that they involve a more closed loop of energy and resources, giving rise to "an industrial economy that is restorative or regenerative by intention and design" (Ellen MacArthur Foundation 2013, 14). As ideology and imagination, ideas about the CE have a wide appeal. On the geopolitical stage, the CE has become an important dimension of policies in China (Zhu et al. 2019), the European Union (Mhatre et al. 2021), and the United States (Schumacher and Forster 2022; Guerra and Leite 2021), promising increased resilience, sustainability, and efficiency in the aftermath of COVID-19 and in the face of growing international tension and environmental concerns. For business, the CE offers an interesting model where waste and surplus

S. Johansson (✉)
Enskede, Sweden

P. Kuppinger (ed.), *Second-Hand Cultures and Economies of Reuse, Repair, Sharing, and Care*,
https://doi.org/10.1007/978-3-031-99875-1_11

materials—traditionally viewed as a cost of production—can become useful and profitable input for other producers (Lüdeke et al. 2019). For grassroots organizations, the CE can become a guiding principle for repairing and reusing consumer products and building social relations, non-consumerism, and new lifestyles (Bradley and Persson 2022). As ideology and imagination, circular economies have many proponents, in various places and at different scales. Some ideas about circular economies might be too idealistic and remain a largely unassailable goal.

In this chapter, I critically re-evaluate ideas of circular economies as inevitably positive processes. To do so, I move away from ideas of the CE as an abstract concept and imagination, which tend to frame the CE as something rational, depoliticized, and universal, often assuming that its reality would play out roughly the same across the world. Following Brenner and Theodore's (2005) intervention about neoliberalism and Shelton et al.'s (2015) contribution about smart cities, this chapter explores examples of existing circular economies, more concretely the example of reclaiming and reusing wood in the city of Detroit. Ideas about circular economies include many positive elements and desirable solutions, but it is important to examine how these ideas are realized and inserted into concrete places. This can include problematic practices that need consideration and attention. Grounding circular economies in the messy complexities of urban life requires an analysis of how they come to interact with other dynamics of urban transformations. In this chapter, I analyze concrete practices and processes of reclaiming wood in the geographic, historical, economic, political, and cultural context of Detroit and how they shape particular patterns of reuse and recycling.

In the context of Detroit's journey from "20th century boom town to poster child of post-industrial decline" (Draus et al. 2014, 2423), the business of reclaiming the wood of the city's many abandoned properties promises to turn a form of waste into a potentially profitable resource. Reclaiming wood sounds like a worthwhile activity that sits well with broader trends of reuse and recycling. It represents an example of the more closed loop of a CE, whereby the discarded, abandoned, or surplus wood is reclaimed and reinserted into market circulation rather than being left to rot or send to the landfill. The fact that Detroit continually ranks as the nations most segregated urban area (Logan and Stults 2011), with a deep and violent history of race and class struggles (Sugrue 1996), embeds the reclamation of wood in multi-layered dynamics and circumstances that add a complex dimension to the recycling process.

In the following, I explore how reclaimed wood in Detroit illustrates broader urban transitions, where Detroit is being reclaimed by a white and wealthier demographic (Johansson 2022). I discuss the link between poor and black residents experiencing foreclosures and evictions, and how these processes create an abundance of raw materials. Simultaneously, the current gentrification of Detroit creates a consumer base for the consumption of reclaimed wood that has been upcycled into artisan furniture and other items. I challenge, complicate, and question some of the "niceness" or positive feelings that are inherent in notions such as upcycling and circular economies. I show how the CE at times relies on the linear economy of "take, make, waste" and ask questions about how to think about "circular loops" within a largely "linear" economy and whether the very material for this "circular loop" is even available in the logic of a circular economy. Can elements be truly "circular" and thus part of positive exchanges when they were sourced under questionable or outright unjust circumstances? Finally, I explore reclaimed wood and products made from it as part of an aesthetic language that decorates the houses of more privileged newcomers to the city, claiming the city's history but taking an interesting distance from the suffering that comes/came with urban decline. Reclaimed wood in its new context of upscale urban homes advertises their connections to the city, symbolizes the flows of people and wealth, and very importantly reclaiming place and belonging in a "new" Detroit.

Joe's Table

Let me begin by introducing Joe,[1] who was living in my neighborhood in Detroit. He had just moved there when I began my research in 2015. He had grown up in a wealthy suburb in northern Detroit but his paternal and maternal grandparents had once lived in the city. They had been part of the exodus of white families that left the city after the "events" of 1967 which white urbanites tended to call "a riot," while Black residents tended to call them a rebellion or uprising. Joe, like his grandparents before him, is part of larger migratory trends of whites in the region. When Joe moved back into the city, Detroit witnessed for the first time in 65 years an increase in its white population (Aguilar and Macdonald 2015). Younger, wealthier, and white individuals, like Joe, have been moving to the city, often from nearby suburban communities. They have established new households using different or "alternative" patterns and styles. This demographic incidentally forms the principal consumer base for products and places made from or with reclaimed wood.

One day, Joe invited me to his house, an extravagant, attractive, and spacious property built in the 1920s, which like neighboring houses had been built to serve as residences for General Motors executives. Joe lived there with his partner. He worked either from home or at a sustainable business incubator in downtown Detroit. The couple's dining room was dominated by a massive wooden table. I complimented Joe on this large and beautiful dining table. My comment made him smile and he explained that this table was custom made by a local Detroit furniture maker, who only employed local residents and used locally sourced reclaimed wood. This company exclusively worked with reclaimed wood, which they sourced from abandoned houses all over the city. Joe's table was made of wood from a house at 1913 Elmhurst Street, its address etched into the corner of the table.

Joe admired the local origin and qualities of the table and was proud that his dollars had supported the local economy that had been in decline for many decades. He enjoyed the table's superb aesthetics and how the different pieces of wood were pressed together into one large slab, yet each individual piece retained a slightly different color. Some pieces were lighter, some darker, and the neat juxtaposition of their unique color schemes, Joe remarked, illustrated the table's "diversity" and history. Joe also pointed to his own special emotional ties to this wood and its original home: in the 1950s, his maternal grandparents had lived on Elmhurst Street, a central reason why he had wanted this particular table. Sitting in his dining room in 2015, the table neatly connected Joe to Detroit's history and his own family's history in the city. Although none in his living family members had any memories of who lived in the house at 1913 Elmhurst when the grandparents had also resided on that street, the table served as a canvas for Joe's imagination. Perhaps his mother and her siblings had opened the door or played games somewhere close to what was now his table's surface?

Joe's dining table not only linked him (and his imagination) to the city's past, but the table also represented ideas and lifestyles of the future. It was a concrete and material manifestation of what Joe conceived of as the CE, an economy that reused energy and materials that would otherwise be discarded as waste. To Joe, wood to be reclaimed—along with other materials—was a resource available in abundance in Detroit. Moving his hand across the table's smooth surface, Joe insisted that this was a more optimal use for the material than ending up in a landfill. He reasoned that trees were once cut down to make the house and furniture at 1913 Elmhurst, and if this wood had not been reclaimed, more trees would have been cut down to make his table. He articulated his purchase of an expensive artisan piece of furniture as a form of caring, help, and support. With the purchase of his table, Joe figured, he helped the environment, supported the local economy, helped preserve both the city's and his family's history, and, very significantly, he outlined and supported an emerging urban future. He understood this future as a more sustainable one, where waste became a "natural" resource, much like a tree. More immediately, Joe had an imposing and stunningly beautiful table that often served as a conversation piece for guests and allowed Joe to lay claims to care, help, and support across its "diverse" surface. The table represented an urban past and simultaneously pointed to the future, urban sustainability, and circular economies. Incidentally, at present, the table also bespoke Joe's economic, social, and cultural position in the city, as a privileged returnee taking advantage of earlier decline while also trying to reverse the tide of exactly this decline.

Reclaiming Wood

Joe's dining table was crafted by a Detroit business called SecondLifeWoods that specializes in making furniture and other smaller items from reclaimed wood. SecondLifeWoods was founded in 2013 and has considerably grown since then. When I began researching the use of reclaimed wood in 2016, the business had just moved into a larger space, hired a dozen new workers, and was in the process of negotiating a contract with a national upscale department store chain. At my first visit to the SecondLifeWoods manufacturing site, the shop floor was busy with people moving things around, processing a vast variety of reclaimed woods into pieces that could be reassembled into new products, all while making sure that they manifested the kind of "diversity" that Joe admired in his table. Jessica, one of the owners of SecondLifeWoods, explained that this was a good time for reclaimed wood. The city's long decline and loss of inhabitants had produced huge supplies of wood hidden away in the seemingly endless numbers of abandoned and disintegrating houses and buildings. Jessica added that most of their customers were from the city's suburbs or even more distant places. Customers who lived in the city tended to be individuals like Joe, who had recently moved or returned to Detroit and exclusively represented higher income groups.

Jessica explained that reclaimed wood had unique qualities that made it an attractive resource. Very crucially, a significant portion of these materials consisted of old-growth hardwood. Jessica doubted that it was even possible to find this kind of wood anywhere these days, since Michigan's old growth forests had been cut down long ago. They were replaced by commercial tree farms that produce wood that is neither as dense, nor as strong and resistant as the naturally grown and aged trees of the past. Another vital quality of the old wood was that it was sourced in Detroit. This provided the resource with a cachet that allowed for premium prices. Owning a piece of Detroit (and its history), Jessica argued, was to possess a piece of concrete local and American legacy, which distinguishes it from wood reclaimed in other cities, even though this resource could have the same material qualities.

Because the abandonment and decay of Detroit are widely known, much discussed, and by now almost symbolic for urban decline, the fate of automobile cities, and rustbelt decline in the United States, the work of SecondLifeWoods gained further value, worth, and recognition. Although Jessica repeatedly stressed that they were a *business* and that they operated for *profit*, consumers and retailers thought that they were also doing good deeds. By "reversing" abandonment and turning it into something beautiful and profitable, they were addressing one the Detroit's most iconic problems: exceptional abandonment and material deterioration.

SecondLifeWoods procured reclaimed wood in different ways. Some materials were donated by individuals who wanted to give them a "second life" because it aligned with their perception of doing good. Other pieces were purchased from a large Detroit retailer of reclaimed wood and then turned into finished products. Another strategy to acquire wood is to purchase an abandoned single-family house and disassemble it. The wood for Joe's table had been sourced in this manner. The method of buying and disassembling a house is expensive, estimated to cost about 10 times more than if the home were bulldozed.

Calculating whether to disassemble a house is difficult in Detroit in 2016 (and partly continues to be so at present). Detroit's depressed real estate market made disassembling houses a more cost-effective alternative than in other cities. However, widespread abandonment and urban blight created their own unique problems. Cheap properties were likely abandoned properties, but if they had been abandoned for a while, scrappers—the more informal and illegal participants in the city's lively housing disassembly industry—had most likely already "visited" a property. Most scrapers are not particularly interested in wood, but remove metals in wiring, plumping, and gas pipes and thus produce damage—through water, cold, or fire—to wood, which greatly affects the quality of resources that can be harvested from a house. The best candidates for disassembly are properties that were relatively well-maintained, which in practice meant that they would be livable, or at least in which someone, until recently, had lived. Accordingly, Jessica had found some of the best sources for sourcing wood at the city's (often sizable) tax foreclosure auctions. In 2015 alone, one in seven Detroiters had been at risk of being foreclosed by the city as its post-bankruptcy-emergency-management severely tightened fiscal control (Hackman 2014), thereby putting an abundance of livable properties on the market at once, further depressing the price of real estate. The best sources for reclaimed wood—maintained properties where someone had lived until not too long ago—were thus far removed from the imagery of urban blight and the effects of decades of abandonment.

Reframing the City (with Wood)

Situating reclaimed wood as an "actually existing circular economy" opens up a problematic side of imaginaries regarding such reclaimed wood in Detroit, uncovering a messier and more ambiguous moral economy. Although reclaimed wood tends to be framed by producers and consumers as a kind of natural and sustainable resource in Detroit, such framings obscure how and why the material has emerged as a resource. Pointing to sustainability and circular economies and using terminologies of reclaiming or recycling diverts the view from pressing social and political issues that are deeply embedded in many of the reclaimed wooden pieces. Representing products like Joe's table in this lingo detaches them from their former use as part of someone's home, someone who might have been forced out, dispossessed through corporate and municipal foreclosures. From working with direct action groups opposing evictions, I know from firsthand experience that residents whose homes were foreclosed were often severely marginalized; many were elderly, black women struggling among other issues with medical problems.

The qualities of the reclaimed wood and its artisanal reworking or upcycling turned the resulting furniture into the hot and trendy products for Detroit's new demographic of predominantly white, wealthy, and younger residents, who were in the process of gentrifying central parts of city. In 2016, reclaimed wood was not only found in the homes of these gentrifiers, but it was increasingly also used in the workspaces and leisure sites that cater to this constituency. Joe's office, for instance, features a wall of reclaimed wood. Some bars, cafés, restaurants where this new urban class gathers use the flair of "retro" reclaimed wood to decorate their premises. When used as part of their external decoration, these establishments discretely signal their social and cultural universe to the city at large.

Situating reclaimed wood against processes of dispossessions, which are often central to its extraction, and processes of gentrification, central to its consumption, illustrates how this material can be a vital part of other forms of circularity and reclamation. The continued influx of white, wealthier, and younger constituencies from the suburbs to the inner city was often called, by both locals and the group itself, a "return." Yet only very few had actually ever lived in the city before. The larger effect of this "return" upon urban space was praised and challenged by observers. As Joe's nostalgic relations toward his family's history illustrates, wood was not the only thing travelling in circles, nor was wood the only thing being reclaimed in Detroit. Given its role in the aesthetic order of an emerging and powerful new class of residents, the process of reclaiming wood in Detroit in intricately interwoven with the reclaiming the city of Detroit.

Concluding Remarks

The analysis of Joe's table provides insight to a complicated landscape of urban decay, gentrification, urban lifestyles, and economies. I showed how complex processes of reclaiming and recycling wood are deeply embedded in urban inequalities. The current CE of reclaimed wood in Detroit is both more complex and more morally ambiguous than the ideas and ideologies it rests upon. In practice, the CE of reclaimed wood does create a market, employment, and a way of transforming one of the most emblematic symbols of post-industrial decline—abandonment—into a profitable resource. Without doubt, this resource is put to better use in new consumer products than as additional landfill material. However, it is important to note that the concrete context of Detroit's wood-reclaiming circular economy, with its recycling, upcycling, and re-circulation of wood in Detroit, is deeply rooted in the particularities that create and give meaning to that space, including its structural and historical inequalities. This circular economy's technical and ostensibly apolitical aim of a closed loop, where waste and resources are successfully removed from the system, must be carefully considered in practice. The case of reclaimed wood in Detroit illustrates that for a circular economy to realize its potential of being truly restorative and regenerative, one must consider where the waste comes from, why it exists as waste in the first place, and what relations of power are used to frame it as a valuable resource and desired commodities. Such an analysis needs to examine how this circular economy is situated in the social reality of its place and time. We need to explore and understand how reclaimed materials have previously been woven into people's lives and their respective places. Otherwise, we do not recognize the problematic nature of the sourcing of these materials. Such an examination makes "admirable" products and production processes appear in a different light. What might look like a worthy recycling and second-hand project takes on more ambivalent and possible problematic characteristics. It is not enough to look only at materials and material processes, or else the alternative offered by a circular economy is limited to being a more efficient and environmentally sound way of reproducing social and economic inequalities. It might be environmentally more sustainable but remains based on largely ignored or unacknowledged social and economic inequalities. As such it cannot be a model for other future projects.

Note

1. All names, including the furniture company, are pseudonyms.

References

Aguilar, L., and C. Macdonald. 2015. Detroit's White Population up After Decades of Decline. *The Detroit News*, September 17, 2015.

Bradley, Karin, and Ola Persson. 2022. Community Repair in the Circular Economy—Fixing More Than Stuff. *Local Environment* 27 (1): 10–11.

Brenner, Neil, and Nik Theodore. 2005. Neoliberalism and the Urban Condition. *City* 9 (1): 101–107.

Draus, P. J., J. Roddy, and A. McDuffie. 2014. We Don't Have No Neighbourhood: Advanced Marginality and Urban Agriculture in Detroit. *Urban Studies* 51 (12): 2523–2538.

Ellen MacArthur Foundation (EMF). 2013. *Towards the Circular Economy*. Vol. 1. Isle of Wight.

Geissdoerfer, Martin, Paulo Savaget, Nancy M. P. Bocken, and Erik Jan Hultink. 2017. The Circular Economy—A New Sustainability Paradigm? *Journal of Cleaner Production* 143 (1): 757–768.

Guerra, Beatriz C., and Fernanda Leite. 2021. Circular Economy in the Construction Industry: An Overview of United States Stakeholders' Awareness, Major Challenges, and Enablers. *Resources, Conservation and Recycling* 170 (1): 105617.

Hackman, Rose. 2014. Detroit's young gentrifiers face a daunting task in buying $500 homes: evicting poor residents. The Guardian, November 27, 2014.

Johansson, S. 2022. *Comeback Detroit: The Return of Whites and Wealth to a Black City.* Dissertation. Department of Anthropology. Stockholm: Stockholms Universitet.

Logan, J. R., and B. Stults. 2011. The Persistence of Segregation in the Metropolis: New Findings from the 2010 Census. *Census Brief Prepared for Project US* 2010.

Lüdeke, F. F., S. Gold, and N. M. P. Bocken. 2019. A Review and Typology of Circular Economy Business Model Patterns. *Journal of Industrial Ecology* 23 (1): 36–61.

Mhatre, Purva, Rohit Panchal, Anju Singh, and Shyam Bibyan. 2021. A Systematic Literature Review on the Circular Economy Initiatives in the European Union. *Sustainable Production and Consumption* 26 (1): 187–202.

Schumacher, K. A., and A. L. Forster. 2022. Textiles in a Circular Economy: An Assessment of the Current Landscape, Challenges, and Opportunities in the United States. *Frontiers in Sustainability* 3 (1): 1038323.

Shelton, Taylor, Matthew Zook, and Alan Wiig. 2015. The "Actually Existing Smart City". *Cambridge Journal of Regions, Economy and Society* 8 (1): 13–25.

Sugrue, Thomas J. 1996. *The Origins of the Urban Crisis: Race and Inequality in Postwar Detroit.* Princeton, New Jersey: Princeton University Press.

Zhu, J., C. Fan, H. Shi, and L. Shi. 2019. Efforts for a Circular Economy in China: A Comprehensive Review of Policies. *Journal of Industrial Ecology* 23 (1): 110–118.

PART IV

Religious Second-Hand Circuits

CHAPTER 12

Second-Hand for the "Third World": Charitable Gift Giving During Religious Volunteer-Tourism in the Dominican Republic

Nicole Nathan

One day while conducting ethnographic fieldwork on American evangelical volunteer-tourism in the Dominican Republic, I noticed a Haitian-Dominican child wearing a t-shirt identical to one that I own, which had been made for my fifth-grade class graduation. A shirt imprinted with my name and those of my classmates, of which maybe a hundred total were made, had somehow ended up almost twenty years later, two thousand miles away in a *batey*, or Caribbean sugar cane plantation. Seeing such a tangible and specific connection to my life outside of fieldwork was somewhat surreal, even though its presence there could be easily explained by my research. In all likelihood, one of my former classmates or someone close to them had brought the shirt to the Dominican Republic to donate second-hand while participating in volunteer work. Still, this encounter was a personally revealing moment of disjuncture for me, in which I was

N. Nathan (✉)
New York, NY, USA
e-mail: nicole.nathan@cuny.edu

P. Kuppinger (ed.), *Second-Hand Cultures and Economies of Reuse, Repair, Sharing, and Care*,
https://doi.org/10.1007/978-3-031-99875-1_12

confronted by how my relationship with my research participants did not start when I first landed in the Dominican Republic to begin my fieldwork. Rather, through the transnational circulation of trade, tourism, and charity, I had always been connected to my research community in myriad, seemingly invisible—but not untraceable—ways.

American volunteer-tourists and the Dominican laborers they seek to aid are enmeshed in pre-existing global market relations. These relations, the results of past and current US foreign policy, late-capitalist modes of governance, and the trade of commodities, particularly sugar and textiles, are part of a global system that has trapped sugar cane and factory workers in poverty. As I will explain, American volunteer-tourists are motivated to alleviate this poverty at the face-to-face level through charitable gift giving, typically of second-hand goods. In this chapter, I examine this cycle of exchange between Dominican-produced commodities to the United States and the counterflow of American second-hand donations to the Dominican Republic. Under late capitalism, charitable giving and commodity exchange are mutually reinforcing systems on both material and emotional levels. While gift and market economies have been historically conceptualized in opposition to each other, more recently, anthropologists (e.g., Bornstein 2005; Goode 2006) have tried to reconcile the perceived dichotomy between the two. I argue that charitable gift giving, particularly of second-hand goods, is embedded in and can even help to reproduce conditions of late capitalism. Notwithstanding the ostensible aims behind it, the charitable giving that takes place during volunteer-tourism—particularly American evangelicals' preference to travel and donate second-hand goods—has many unintended and sometimes harmful consequences on the recipient community.

Each year, an estimated four million Americans travel abroad as participants in short-term missions (STMs), colloquially known as mission trips, the religious branch of the rapidly growing volunteer-tourism industry (Occhipinti and Priest 2014). Roughly 24 percent of all American evangelicals have participated in short-term religious service projects (Smith and Snell 2009). Volunteer-tourism is designed around engagement with local communities and humanitarian relief. As a faith-based subset of volunteer-tourism, STMs promise an additional religious purpose: evangelization, a key tenet of evangelical Christianity to transform the lives of others through religious conversion. For these reasons, STMs are an important part of charitable giving and religious socialization in the United States. STM aid is provided in the form of construction projects, medical clinics, and donated goods and money. For purposes of this chapter, I focus on second-hand goods, including why evangelicals prefer to give second-hand goods and how such goods may provide some temporary relief, but ultimately are part of the system that necessitates their donation in the first place.

Proponents promote volunteer-tourism as a potentially transformative experience for Americans. Building on the work of Howell and Dorr (2007), I examine STMs as postcolonial pilgrimages centered around Turner's *communitas*, or sense of community based in shared experience (Howell 2012). STMs become sanctified through the acts of seeing and serving the poor. During these dynamic encounters, meanings and identities are forged between actors who have "different cultural backgrounds and unequally positioned stakes in their relationships" (Faier and Rofel 2014, 364). Significantly, STMs expose American volunteers firsthand to vast global and racial inequalities. In this way, STMs can powerfully (re) shape how participants conceptualize and understand culture, globalization, race, and economic inequality—in particular, the ways in which they reflect on or confront how they are implicated in others' economic positions.

My research on STMs focuses on the Dominican Republic, one of the more popular American STM destinations due to its geographic proximity to the United States, its presence of established Christian churches with whom they can partner, and its popularity as a traditional tourist destination (Priest and Priest 2008; Priest et al. 2010). The major presence of the tourism sector, as well as other dominant industries like agriculture and textile, on one hand contribute to the Dominican Republic's GDP, making the nation an upper-middle income developing country. However, these industries also create many adverse social and economic conditions that attract volunteer-tourists to the region, including extreme inequality and labor abuses.

The STMs that I study take place in the factory town and *bateyes* of "Azúcar Mayor," not far from some of the country's most popular tourist destinations. Azúcar Mayor is one of the largest privately owned sugar producers in the Dominican Republic and a sugar supplier for several major multinational brands. The company owns a luxury resort where STM "teams" sometimes stay during their time in the country. Azúcar Mayor's sugar-cane plantations and refinery factory employ many of the province's residents. Many of these workers are undocumented Haitians or of Haitian-descent, and because of their undocumented status face great difficulty finding formal employment outside of the sugar industry, where they work under hard and dangerous conditions for very low wages. As tourist encounters at sites of labor, STMs make visible the oppressive working conditions that have been denounced by human-rights organizations as exploitative and even coerced labor. However, too often these conditions are decontextualized during STMs from the broader sociohistorical context and Americans' own positioning to the problem. American STM teams are complicit, not only through their consumption of sugar and textiles produced in the country, but also because their methods of giving, in conflict with their intent, in some ways help to sustain the whole system. It is no coincidence that contemporary life is marked by historic rates of voluntarism and philanthropy yet ever-deepening wealth inequality. The paradox of philanthropy and charity is that they are made possible by policies that are a root cause of economic inequality and social injustice.

Evangelical Giving Ideologies, Gift Exchange, and Second-Hand Goods

Giving to the less fortunate is a long-valued Christian practice. Christian giving is influenced by Christ's practice of charity as well as tenets expressed in the Bible, such as compassion and mercy. There are many models for charitable giving in Christianity, such as the practice of tithing, or giving one-tenth of one's income to the church. Religious-motivated giving is the most common form of charity in the United States, where 73 percent of charitable giving goes to organizations with religious ties; the largest share of charitable donations come from members of conservative Christian groups (McKitrick et al. 2013).

From their daily religious devotionals to the names of their missions, mission volunteers frequently reference giving narratives from the Bible as influential to their own giving behaviors. For instance, one team refers to themselves as the Good Samaritans, a name inspired by the eponymous parable in Luke 10:25-37. Epitomized by a foundational value of Christianity, "Love your neighbor as yourself" (New International Version 2024), the moral lesson of the Good Samaritan emphasizes providing for anyone in need, regardless of their cultural or religious background. Aid that transcends difference is a stated secondary goal of American volunteer-tourists. Their primary aim is to answer Christ's call in Mark 16:15 to "go into all the world" and evangelize.

Evangelical volunteers' giving behaviors are guided by religious beliefs. One prominent giving ideology is the preference among short-term missionaries to donate in-kind used goods, which are often ineffective and costly to ship, rather than donating cash or buying the same kinds of goods new in the Dominican Republic. One reason American evangelicals prefer giving second-hand items over cash is because it often serves to satisfy the giver's sense of moral responsibility, alongside other personal benefits. Their aims are less about environmental consciousness and more about constructing a moral subjectivity. Through performance of moral dispositions, like giving away their own possessions and proximity to Dominicans' poverty, STM volunteers cultivate a moral subjectivity. Short-term missionizers travel to donate goods themselves so that they can temporarily move away from the materialistic trappings of the modern world to learn moral lessons and be transformed by the poverty that they witness.

Evangelicals follow Christ as the paragon of giving, and in Matthew 5:40 he instructs his followers to give their coat to those who ask for it, or essentially give the shirt off one's own back. In one notable example of this, Michael, an American STM participant, purchased a brand new, high-quality wardrobe to wear during his mission that he then proceeded to gift at the end of his trip, returning home empty-handed. Many STM volunteers seek to emulate a more moral life and even to live in a "Christ-like" manner. Performing manual labor themselves and donating the shirt off their back instead of hiring a professional and sending money are constructed as a sacrifice and a way to serve God authentically.

The circulation of one's own second-hand goods, more so than of money or newly purchased commodities, fulfills Marcel Mauss' seminal theory on gift exchange in which to give something is to give a part of oneself. In his seminal essay *The Gift*, Mauss (2000) asserts that gifts are part of a system of recognition in which social solidarity is achieved. The mutual obligation to give, receive, and reciprocate forges an irrevocable bond among participants. Mauss contends that gifts are imbued with the spirit of the giver, and this quality remains with the object as it is exchanged. To give a gift is to give a part of oneself. Giving second-hand clothing is a salient example of this because clothes are a powerful tool for expressing identity through personal taste—the costume in the performance of religion, ethnicity, class, and gender. If clothes are an extension of the self, then sharing one's clothes constitutes a form of recognition that creates an appearance of egalitarian relations. One of the most common practices of giving clothes second-hand is the hand-me-down or reuse of clothes and other items among family, typically passed down from older siblings to younger ones. Giving one's own clothes, as Michael did, can reflect the religious kinship between Americans and Dominicans as "brothers and sisters in Christ" and remind the recipient of their relationship every time the item is worn.

In contrast to Michael's like-new clothing, more often, the second-hand goods that donors give are already starting to see some wear—or even damage—and are no longer considered good enough or of use to the American donors. These rejected goods reinforce the power differences between donors and recipients. Discarded, undesirable second-hand goods effectively diminish the recipients to the second class. While the second-hand gift is imbued with the essence of the giver, its damage is a physical marker of the inequality between them. For this reason, charitable giving has historically been omitted from Maussian gift exchange theory because charitable giving occurs between unequal parties, which impedes reciprocation, and thus precludes solidarity. As Douglas wrote in her foreword to *The Gift*, "Though we laud charity as a Christian virtue we know that it wounds" (1990, ix).

Mauss confronts the notion of the pure gift, or altruism, by demonstrating that gifts can never be free of an ulterior, self-serving motive. The gift and the act of giving are inextricably connected to the giver. In a similar manner, evangelical giving ideologies sanctify the good intention of the gift giver over the quality or usefulness of the gift itself. For instance, Acts 20:35 is often referenced by volunteers: "It is more blessed to give than to receive." American evangelicals discursively construct their donations as a way to "bless" Dominicans, assigning a moral, even divine agency to themselves. As a way to cultivate moral and spiritual capital, the donation of used goods can benefit the giver more than the recipient, especially when the focus is placed on the ritual of gift giving and the bond between the parties over the usefulness of the gift itself. In some cases, charitable giving is more about donors feeling good than actually doing good.

By donating their own possessions second-hand, Americans have the additional moral benefit of freeing themselves from their worldly possessions. For evangelicals, living in a Christ-like manner can demand abstaining from materialism. Many evangelicals believe that the privileges of their relative wealth can be detrimental to their spiritual life. This belief is rooted in another Bible verse that is commonly brought up during STMs, Matthew 19:24: "It is easier for a camel to go through the eye of a needle than for someone who is rich to enter the kingdom of God." STM volunteers also reference Christ's declaration in Luke 6:20, "Blessed are you who are poor, for yours is the kingdom of God."[1] American evangelicals are thus motivated to participate and to serve in STMs. In bearing witness to Dominican poverty, American volunteers gain insights that promote a life not centered around materialism but instead on relationships. They atone for their consumerist behaviors back home by turning away from spending money on material things and giving away their own possessions to the poor. Reusing goods by passing them on second-hand, rather than throwing them away, serves as a moral way to discard unwanted or even damaged goods. Wastefulness is considered sinful. In giving away their possessions, Americans can strengthen their relationship with fellow Christians and with God.

Regardless of the condition or usefulness of the second-hand item, volunteers express expectations for the ideal recipient, including demonstration of gratitude and joy. For instance, as Maggie, an American participant, writes in her journal, "I think about the *bateyes*[,] and how on one hand, people seemed so desperate that they would take anything[,] but yet on the other hand, they were so proud, they wanted a choice of clothing to be given. One extreme to the next." Maggie's attitude reflects common cultural tropes and myths of charity in the United States and reproduces the adage that beggars cannot be choosers. Teams sometimes distribute used clothing randomly, without concern for the size of clothing. On one occasion, a Haitian-Dominican woman was given some skirts, and when she asked through an interpreter for pants because she did not wear skirts, her request was denied. This interaction was recounted in shock many times by the team because it violated their cultural rules of gift exchange. It was moreover perceived as a spiritual rejection because the gift was imbued with the spirit of the giver and because it was a rejection of their religious mission. The skirt exchange demonstrates the ways that Dominicans are not powerless in one-sided charitable giving and instead play a complex agentive role by taking up Americans' gift-giving ideologies and sometimes (re)producing or challenging them. The woman's perceived lack of gratitude is especially powerful in the charitable giving context because it leaves the charitable exchange unfinished, denying the giver feelings of altruism. While lack of gratitude can indicate a form of everyday resistance, in this case, it criticizes the ineffectiveness of the gift. However, rather than causing the team to reflect on how to better provide donations to recipients, the Haitian-Dominican woman was dismissed by the team as rude and ungrateful.

The (Second-Hand) Goods, Bad, and the Ugly

The circulation and reuse of second-hand goods is more environmentally sustainable and cost-effective by reducing the manufacture and distribution of new items. However, in-kind donations have been widely criticized across aid research because they are frequently politicized and sometimes serve the interests of the donors at the expense of the recipients (Clapp 2005). Cash donations are almost always preferred by recipients and recommended by international aid organizations because it boosts local economies and provides greater flexibility for the recipients. Recipients of cash are empowered with the agency to decide for themselves what they need, and the purchase of new goods in the recipient community helps stimulate the local economy. In contrast, charitable giving of second-hand goods can diminish the need for products to be manufactured and sold locally. Imported in-kind donations drive Dominicans out of work and destroy entire industries, making Americans complicit in the poverty that they are trying to alleviate.

The donation of second-hand goods during volunteer-tourism is frequently misguided, self-serving, and even sometimes harmful. One common issue is donation dumping, or the excess donation of unusable goods that are better suited to be recycled or thrown away. Donation dumping has rapidly increased alongside the mass production and overconsumption brought on by the rise of fast fashion. Donations of second-hand goods to the Dominican Republic are typically made based on little to no consultation with the recipient community and are sent without regard to actual needs or wants. Such was the case when American volunteers at a mission-run Dominican school donated their used textbooks. The textbooks were written in English and many were on subject matter specific to the United States, such as US history. The donation was useless to the Spanish-speaking Dominican school. Besides being an ineffective gift, donation dumping burdens the recipients who must store and spend time sorting through and throwing away unusable items at their own expense. Donation dumping devalues the recipient who is presumed to accept anything, regardless of how broken or dirty. The recipient takes on the sins of materialism and wastefulness through the second-hand goods that Americans are keen to discard but do not want to be burdened by the guilt of their disposal.

Donation dumping requires more work and can sometimes cost more than purchasing goods new. For example, an oversized box of hundreds of misprinted t-shirts were brought by Joe, a mission leader, and his church. The box sat in customs limbo for over a year after airport officials charged Joe an unexpected duty for importing what they deemed commercial goods. Its release was secured only after Joe contacted his US senator, who reached out to the Dominican ambassador. As Joe wrote in an update to his church:

> While we are frustrated and do not understand why the airport insists on imposing a tax on humanitarian aid we have brought for their country, we continue to pray that God will move their hearts to compassion (or at least common sense!) and release the shirts so that we can distribute them to the impoverished children of their country.

While Joe and his church perceived there to be a great need for these t-shirts, most Dominicans have access to affordable clothing. Second-hand clothing from the United States is shipped in bulk to other countries, like the Dominican Republic, and sold at low costs in resale markets. While many of Joe's t-shirts were distributed and are worn on occasion, almost a decade later, hundreds still take up space in the church's storage, where they are picked up by the church youth group and repurposed to create costumes for their church performances.

Misguided gifts and methods of giving place unnecessary financial burdens on recipients to receive the unwanted goods. For instance, an American couple, Ed and Nancy, shipped two large boxes via FedEx to Javier, a Dominican doctor whom they had met through a mission. When Javier collected the boxes, he was charged import duties that equated to 80 USD, which he paid for out of pocket. Initially, Javier was unbothered to spend such a large sum of his own money, thinking that the boxes contained medications for his mission-funded clinic. When he opened the boxes, he discovered that both were filled with old, deflated soccer balls. Six months later, Ed and Nancy came to the Dominican Republic and the soccer balls were still sitting deflated in his apartment since he did not have an air pump nor any use for them. The American couple proposed sending Javier another bulk shipment: this time, hundreds of old file folders from Ed's workplace. Javier politely thanked them but explained that he did not have a need for so many file folders. After ignoring Javier's objections, at last, he revealed the personal cost of accepting their last shipment. Without acknowledging Javier's financial burden or mention of reimbursing him, Ed and Nancy asserted that Javier had incorrectly retrieved the items.

Unfamiliar with the Dominican Republic's import tax to receive goods, they remained determined to ship the folders to the mission at Javier's expense.

The t-shirts, soccer balls, and file folders were all misguided donations that did not take local needs into account, largely because of miscommunication, lack of knowledge of local needs and regulations, and, above all, unwillingness to listen to, and to learn from, Dominicans. Second-hand goods are often left unused and cost the recipients' time, space, and money. More significantly, donated goods from abroad undercut local businesses. In Joe's mission update, he implied that the country should be grateful to receive this donation for its "impoverished children." He decried the tax on this gift without recognizing that such in-kind donations are harmful to local economies. The duty tax serves to offset the negative impact that the shirts have on the Dominican textile and apparel industry, which has been negatively impacted by the CAFTA-DR, the Central America-Dominican Republic-United States Free Trade Agreement. Following the signing of CAFTA-DR, the Dominican Republic faced an overall decline in apparel and textile exports, despite promises that the agreement would create new jobs in apparel factories for export to the United States.

The effects of in-kind donations are one example of what's termed the perverse economics of volunteer-tourism, wherein relatively wealthy people pay money to import their own unskilled labor to poor areas with high unemployment. Robert Wuthnow estimates that Americans spend an estimated minimum of 1.6 billion dollars each year on STMs, money that he points out could sustain the lives of more than four million of the world's poorest people for a year (Wuthnow 2009). Despite these high costs, in my research, the funds contributed toward the aid itself are typically the smallest portion of the overall STM cost; following airfare, lodging, food, and in-country transportation, which all represent greater expenses. If volunteers' goal is to alleviate poverty, STMs are extremely cost inefficient. In addition to aid dispersal, many American volunteers perform construction work that could instead be hired out and employ skilled Dominicans, who would provide a higher quality of workmanship and do so at a fraction of the volunteers' overall cost. For example, Beek and Alan's (2006) study of volunteer-tourists rebuilding homes in Honduras after Hurricane Mitch found that Americans spent an average of $30,000 per house (when including their own travel and lodging costs), while the total for similar construction work by local organizations was a mere $2,000 per house.

For the same amount of money that it cost them to travel to Mexico and build one house, volunteer groups could have built 15 houses and created jobs that would help stimulate the local economy. Importing their volunteer labor and used donations compounds the problems that they profess to address. Cash donations better enable local people to develop their own communities and would have farther-reaching and longer-lasting effects.

Poverty, Moral Subjectivity, and the Aversion to Cash Donations

The preference among American evangelical volunteer-tourists for second-hand donations reflects religious giving ideologies. However, this preference is also motivated by American evangelicals' general reluctance to give cash to Dominicans, even though cash donations would provide better quality goods, stimulate the local economy, and empower recipients to decide what they need most. The reluctance to give cash in part reflects deeply rooted colonialist and race-based ethnocentric beliefs that poor people (and, by extension, poor nations) are fiscally irresponsible and corrupt or dishonest with money. American volunteers and missionaries often consider it necessary to decide what they think is best for Dominicans, rather than being receptive to Dominicans' ideas and their ability to make such decisions for themselves. While gift exchange promotes social solidarity, ideologies of charitable gift giving are fraught with issues of power.

American evangelicals do provide money in consultation with the missionary to select trusted individuals. Money is given to Dominicans through means of employment in the mission, scholarships, and sponsorships. However, STM teams are otherwise often unwilling to give Dominicans money due to mistrust. Missionaries tell cautionary tales about Dominicans taking advantage of their generosity. One such story told often by Jason, a long-term missionary based in the country, recounts an American man who, over the course of a week, gave thousands of dollars to a Dominican man unknown to the mission. The cash donations started with a small sum when the man asked for help buying food for his family, but he then requested and received increasingly larger sums including money to buy a motorcycle. The missionary offered no proof that this was a scam, but it was offered as a fact. Due to this fear, American evangelicals feel more comfortable giving money to the long-term American missionaries to manage as they see fit. However, the misuse, mismanagement, and even on occasion misleading use of funds by American missionaries have cost short-term teams far greater sums of money[2] than any potential Dominican "scam."

Americans frequently express the misguided concern that giving cash will foster what they call a culture of dependency, wherein charity destroys the recipients' personal initiative and erodes their work ethic. They fear cash donations will create a permanent underclass. As such, Dominicans who ask for cash outright may be castigated as greedy and undeserving. One American group compared what they called poor Dominicans' "gimme attitude" to their understandings of poverty in the United States. According to one woman, Kelsey, "It's the ones who don't work and don't go to church who want [us to give them money], and it's like that in the US too with welfare. It's a culture-of-poverty because it is easier [for them] to just stay home and get a check." Dominican spirituality and poverty are both paradoxically described by American evangelicals. Albeit a myth, Americans consider the culture-of-poverty to be sinful. At the same time, Dominicans' poverty is in other ways valorized as a life condition that is thought to facilitate the pursuit of a more moral or spiritual life. According to recent college graduate Dwayne, "[Dominicans] might have nothing but a smile on their face. No matter how little they think they have, they have joy, and they don't know what they don't have. They have the Lord and they know that." Americans believe that Dominicans' poverty is sanctified and predisposes them to be less materialistic, which fosters spirituality that would be tainted by cash. This belief is held simultaneously, and without perceived contradiction with the belief that Dominicans' poverty is the result of irrational economic behaviors, which are construed as immoral.

The mistrust in Dominicans' ability to demonstrate personal responsibility is expressed by Americans through neoliberal tropes that pathologize poor people as lacking self-control. As Jason, the missionary, tells visiting teams:

> Dominicans will get $800 and they won't save it or buy food. They'll go and buy an $800 TV, and I'm thinking, "You don't need an $800 TV when your children don't have food in their bellies." And then they'll need food or something else, so they'll turn around and sell that TV for $200. Well, that's $600 they just lost. And they do this again and again.

These discourses portray poverty as an individual failing, in this case a cultural failing. There are several narratives that criticize Dominicans' purchase of Air Jordan and other luxury sneakers, an emblem of irresponsible Black consumption among white Americans. By foregrounding this emblem, these narratives imbue the class-based concerns of the culture-of-poverty thesis with racializing assumptions. These discourses contribute to the already existing negative perceptions of Black culture among white evangelicals, particularly the perception that "black communities lack strong commitments to the virtues of accountability and fiscal solvency. For white evangelicals, this is more than a practical problem; it is viewed as a distinctly moral and spiritual one" (Elisha 2010, 280). Much like in Omri Elisha's study on partnerships between African American and white evangelical churches, Dominicans are admired by American evangelicals for their perceived deep—and even innate—spirituality. Despite their heightened spirituality, Dominicans, even Dominican pastors, are rarely trusted or given decision-making power over mission donations.

There are STM teams whose leaders and members are more receptive to listening to the community and adapting their methods. For instance, although Joe's misguided t-shirt donation was his own idea, his mission was started after talking to Dominicans about their needs. After learning that the public schools required uniforms, thus making the otherwise free education cost-prohibitive for many Dominican families, Joe's team began taking schoolchildren back-to-school shopping. Students are given a budget to buy their uniforms and school supplies. The children look forward to Joe's mission the most every year because it gives them the opportunity to choose their own new items. However, their purchases are then open to scrutiny by some of the members of Joe's team, who admonish them for not being more resourceful or wise with their purchases. For example, Kelsey criticized the tight-fitting shirts the children selected, which she dismissed as an immodest and trendy style. She wanted to restrict the children's purchases to larger sizes that the children could grow into and would thus last them for multiple school years or be passed down to younger siblings. Kelsey disapproved when she learned that all the children receive a new uniform every year, stating that she frequently wore used clothing as a child, and implying that Dominicans should do the same. By interpreting Dominicans' consumption behaviors according to her own, Kelsey echoes the culture-of-poverty thesis and depicts Dominicans as irresponsible with money and less enterprising than Americans. However, in the Dominican Republic tight-fitted clothing is considered more orderly and even required in school. Additionally, because Joe's mission only provides the children with just one shirt, pair of pants, and pair of shoes, the uniforms are mostly threadbare by the end of the school year and are not in a condition to be re-worn the next year. Joe, whose mission had been purchasing uniforms for over a decade, was more aware of these challenges and corrected his team member.

Evidence supports the effectiveness of cash donations given directly to recipients, who can make the best decisions on how to spend it for themselves. For example, a Haitian-Dominican woman, María, used a microloan to purchase a solar panel. Although the team initially disapproved of the panel, which they viewed as a luxury item, it proved to be a success, providing a much-needed renewable-power source for her *batey*. María recharges electronics for a small fee and has since paid back her loan and continues to provide for her family. Despite examples like María, many of the American volunteer-tourists feel that they know what is best for Dominicans. Yet their second-hand gifts, as demonstrated in the previous section, are frequently misguided, costly, and go unused because of their lack of cultural knowledge, professional expertise and, above all, their unwillingness to listen to, and to learn from, local input on how to best serve their needs.

The COVID-19 pandemic brought many changes to the volunteer-tourism industry, largely due to travel restrictions. With many Dominicans out of work due to lockdowns and the pandemic's devastating impact on the tourism industry, American churches sent cash donations instead of volunteers. The Dominican pastors through their sophisticated organizing power quickly mobilized an extensive network of their own volunteers to purchase and distribute food aid throughout the region, defying any misgivings about their fiscal responsibility or competency. This method of giving appears to have been short-lived, as the traditional mission model of volunteering and second-hand donations returned when international travel made a comeback in 2022.

Charitable Giving Under Late Capitalism

In this chapter, I have tried to elucidate some of the reasons why evangelical Christians typically—but not always—favor charitable giving of second-hand goods. According to Jason, the long-term missionary, "a lot of people have good intentions but not direction." Lacking relevant professional expertise, American volunteers are not able to address the complex issues that perpetuate inequality. Notably, American volunteers linguistically erase "inequality" and instead focus on what they refer to as "Third World poverty." By erasing notions of inequality, volunteers are limited in the ways that they act—donating second-hand goods that make marginal improvements in material living conditions without addressing root causes. Because volunteers are told that aid changes lives, they "feel good" and do not seek out alternatives that could create more substantive change. Nor do they seek out alternatives to capitalism, consumerism, and inequality back home. Charitable giving and voluntarism do not address the underlying causes of the problems that they aim to solve. This is not to say that STMs do not help; many recipients do benefit from second-hand goods and other aid and services provided by volunteers. Moreover, their second-hand goods find a new life through reuse, avoiding landfills and lessening the demand and environmental consequences of producing new goods.

At the same time, donating used goods provides a moral framework for Americans to continue to justify their consumption of products. Like families in Rhoda Halperin's research (Halperin 2009) on extreme gifting in the United States, when volunteer-tourists return home, they continue practicing unethical consumption. In some ways volunteers may be compelled to consume *more* rather than less because they believe that their materialism has been mitigated by giving away some of their possessions for reuse. More than one American rationalized to me that donating their old clothes made them feel good and provided them with an excuse to update their wardrobe and go shopping for new clothes when they returned home.

Dominican recipients of second-hand goods are expected to be grateful for worn items and engage in morally charged consumption habits. On the other hand, since they can financially afford it, Americans are not considered irresponsible if they overconsume, even though that overconsumption negatively impacts the environment. Although the environment is not evangelicals' primary motivation for donating second-hand goods, any sustainability achieved through second-hand donations is counteracted when such donations spur higher consumption of new products. While the Global North is responsible for the majority of greenhouse gas emissions that contribute to climate change, they have exported the harms disproportionately to the Global South, who have fewer resources to adapt to or respond to extreme weather conditions. It is also the Global South who is burdened with mitigating climate change through curbing *their* emissions levels, not unlike the ways Americans cast off broken second-hand goods to Dominicans who are burdened with storing, sorting, and throwing away the items at their expense.

Providing goods and services to Dominicans also provides a moral framework for Americans to continue to justify their high consumption with little regard of the treatment of workers. Despite having been exposed during the STM to the exploitative conditions of labor firsthand, American volunteers do not seek out fairer alternatives to mass-market sugar or clothing that would pay workers fair wages and ensure safer working conditions. In this modern-day Kula ring, gifts and commodities circulate in a multidirectional flow. Americans donate second-hand clothes to the Global South, only to return home and buy new clothing imported to the Global North, clothing they may one day donate. Arjun Appadurai's work in *The Social Life of Things* (2003) shows how commodities and gifts do not necessarily belong to discrete spheres of exchange. The modes of production of sugar and textiles as commodities, which rely on the State and global markets, influence the movement of people and second-hand gifts through volunteer-tourism. Sugar and clothing can be bought cheaply at American stores, but it comes at a high price for sugar and textile workers who must rely on the charitable gifts provided by volunteer-tourists. These gifts are indirect compensation, supplementing workers' low wages that do not adequately provide for their basic needs. Charitable gift giving under late capitalism is voluntary, at least for the individual volunteer; but charitable giving has in part become obligatory to ensuring the smooth running of late-capitalist production.

Charitable giving in the context of volunteer-tourism in the Dominican Republic attends to participants' deeper, pre-existing relations of commodity exchange. Although Mauss omits charity from his theory on gift exchange, his work provides an important lens to charitable giving when reciprocation is reconceptualized to reflect economic relations under late capitalism. Under late capitalism, the lines between gift and market economies are blurred due to deregulation and retrenchment and privatization of social services, and, as a result, reciprocation can take on other forms. The socially binding forces behind any exchange—gift or commodity—are powered by the need for mutual recognition (Robbins 2009). Wages are supposed to recognize agency and work. When wages are not fair, workers are not fully recognized and devalued in the economic sphere. Charity and social services that supplement wages are disempowering because they recognize only one's class position. While the purchase of Dominican sugars and textiles makes Americans complicit, participation in volunteer-tourism can expound the issue. Knowing their needs will be met by the influx of foreign volunteers and donations, Azúcar Mayor is incentivized to continue to inadequately compensate and treat their workers. What's more, Azúcar Mayor even profits from volunteer-tourism. As mentioned earlier, some volunteer-tourists stay at the Azúcar Mayor-owned resort during the STM. In recent years, American volunteers have expanded their mission to include the resort workers as well, essentially paying the corporation to provide charitable goods and services to its own workers.

In deregulated capitalist societies, the moral concerns of the economy are in flux between society and the state; corporations and the individual are emphasized as responsible for providing for social needs. In the context of a global market economy, charitable giving brings to mind Annette Weiner's theory "keeping while giving" (1992). Americans can "keep" the capitalist system from which they benefit, by giving their surplus, or their used goods, cash, and services. The low price of sugar, clothing, and tourism can be maintained and guilt-free. As Douglas argues, "There are no free gifts; gift cycles engage persons in permanent commitments that articulate the dominant institutions" (1990, xii). Notwithstanding the ostensible aims behind it, the charitable giving that takes place during volunteer-tourism ultimately can help to reinforce oppressive systems of commodity exchange under late capitalism. By meeting some of the workers' basic needs, volunteers work against more permanent solutions by enabling the sugar corporations to evade workers' demands for fair wages and access to medical care.

Notes

1. Significantly, the verse cited here from the Book of Luke is altered in the Book of Matthew, which states "blessed be the poor in spirit." According to some Bible scholars, the phrase "poor in spirit" signifies poverty as both a physical and spiritual condition.
2. One missionary attempted to sell an American mission team land on which the team had already built a medical clinic with their own funds and labor; his asking price included the value of that building, upward of hundreds of thousands of dollars. The team ultimately walked away, purchasing a lot of their own on which to rebuild.

References

Appadurai, Arjun, ed. 2003. *The Social Life of Things: Commodities in Cultural Perspective*. Cambridge: Cambridge University Press.

Beek, Ver, and Kurt Alan. 2006. The Impact of Short-Term Missions: A Case Study of House Construction in Honduras After Hurricane Mitch. *Missiology: An International Review* 34 (4): 477–496.

Bornstein, Erica. 2005. *The Spirit of Development Protestant NGOs, Morality, and Economics in Zimbabwe*. Stanford, CA: Stanford University Press.

Clapp, Jennifer. 2005. The Political Economy of Food Aid in an Era of Agricultural Biotechnology. *Global Governance* 11 (4): 467–485.

Douglas, Mary. 1990. Foreword: No Free Gifts. In *The Gift: Forms and Functions of Exchange in Archaic Societies*, ed. Marcel Mauss, vii–xviii. New York: W. W. Norton & Company.

Elisha, Omri. 2010. "Evangelical Megachurches and the Christianization of Civil Society: An Ethnographic Case Study." In Politics and Partnerships: The Role of Voluntary Associations in America's Political Past and Present, edited by Elizabeth S. Clemens and Doug Guthrie, 269-296. Chicago: University of Chicago Press.

Faier, Lieba, and Lisa Rofel. 2014. Ethnographies of Encounter. *Annual Review of Anthropology* 43:363–377.

Goode, Judith. 2006. Faith-based Organizations in Philadelphia: Neoliberal Ideology and the Decline of Political Activism. *Urban Anthropology* 35 (2): 203–236.

Halperin, Rhoda. 2009. Extreme Gifting: The Moral Economy of a Community School. In *Economics and Morality: Anthropological Approaches*, ed. Katherine E. Browne and Lynne Milgram, 101–122. Lanham, MD: Altamira Press.

Howell, Brian. 2012. *Short-Term Mission: An Ethnography of Christian Travel Narrative and Experience*. Downers Grove, IL: IVP Academic.

Howell, Brian, and Rachel Dorr. 2007. Evangelical Pilgrimage: The Discourse of Short-Term Missions. *Journal of Communication and Religion* 30 (2): 236–265.

Mauss, Marcel. 2000. *The Gift: Forms and Functions of Exchange in Archaic Societies*. New York: W. W. Norton & Company. First published 1950.

McKitrick, Melanie A., J. Shawn Landres, Mark Ottoni-Wilhelm, and Amir D. Hayat. 2013. *Connected to Give: Faith Communities*. Los Angeles: Jumpstart.

New International Version. 2024. *Bible Gateway*. www.biblegateway.com. Accessed 20 February 2024.

Occhipinti, Laurie, and Robert Priest. 2014. *Making a Difference in a Globalized World: Short-Term Missions that Work*. Lanham, MD: Rowman and Littlefield.

Priest, Robert J., and Joseph Paul Priest. 2008. They See Everything, and Understand Nothing: Short Term Mission and Service Learning. *Missiology: An International Review* 36 (1): 53–73.

Priest, Robert J., Douglas Wilson, and Adelle Johnson. 2010. U.S. Megachurches and New Patterns of Global Mission. *International Bulletin of Missionary Research* 34 (2): 97–104.

Robbins, Joel. 2009. Rethinking Gifts and Commodities: Reciprocity, Recognition, and the Morality of Exchange. In *Economics and Morality: Anthropological Approaches*, ed. Katherine E. Browne and Lynne Milgram, 43–58. Lanham, MD: Altamira Press.

Smith, Christian, and Patricia Snell. 2009. *Souls in Transition: The Religious and Spiritual Lives of Emerging Adults.* New York: Oxford University Press.

Weiner, Annette. 1992. Inalienable Possessions: The Paradox of Keeping-while-Giving. Berkeley, CA: University of California Press.

Wuthnow, Robert. 2009. *Boundless Faith: The Global Outreach of American Churches.* Berkeley, CA: University of California Press.

CHAPTER 13

Secondhand Sacred: Christian Material Culture and Reseller Ethics

James S. Bielo

In April 2023, a story about selling Christian material culture on the secondhand market went viral. The story centered a grandchild of Charles Stanley, a celebrity evangelical preacher and author. In the days following Stanley's death, family and friends were invited to go through his home and select memorial keepsakes. A few days later, Stanley's 29-year-old grandson, Matt Broderson, posted a video to his YouTube channel announcing that he would auction two of his grandfather's items on eBay.

The video received immediate jeers. Commenters accused the decision to auction as "disrespectful," "disgusting," and "dishonorable." Christian and secular news outlets seized on the story, asking if the auction was justifiable or in "poor taste."[1] Broderson acknowledged his critics, but let the bidding proceed. On May 1, both items sold. A 13x13 inch autographed portrait of the preacher, dated 1996, received 40 bids and sold for $3650 USD. A pocket watch engraved with the preacher's initials received 63 bids and sold for $2375. The saga resumed a week later when reports circulated that an anonymous buyer returned the items along with an

J. S. Bielo (✉)
Oxford, CT, USA
e-mail: james.bielo@northwestern.edu

P. Kuppinger (ed.), *Second-Hand Cultures and Economies of Reuse, Repair, Sharing, and Care*,
https://doi.org/10.1007/978-3-031-99875-1_13

impassioned plea for Broderson to keep them as family heirlooms.[2] Broderson complied, issuing a public apology.

Stanley's celebrity made this a story and the auction's posthumous immediacy made it all the more appetizing. Still, the saga of Stanley's stuff raises questions of broader relevance that are present every day for secondhand sellers. When do the goods of divestment outweigh the goods of inheritance? Is it ok to profit from materials culturally coded as sacred? Are there ethical limits to this kind of capitalist activity? If so, how are such determinations authorized? Any hesitation toward commodifying the sacred persists against the grain of a long history of selling devotional objects (McDannell 1995; Wharton 2006), whole industries dedicated to selling religious materials (Cantwell 2017; Vaca 2019), and a secondhand marketplace that, on any given day, includes thousands of sacred items for sale in thrift stores, antique shops, eBay, and other venues (Cuperus 2019).

This chapter explores the ethical dynamics of sacred commodification from the ethnographic vantage point of resellers, individuals who scavenge secondhand venues in search of donated, discarded, and passed over items.[3] Some resellers source and sell as a "side hustle" to supplement other income streams while others resell full-time, circulating a greater mass of materials to sustain a livelihood. Some source only to sell, while others curate personal collections that exist in tandem with inventories for sale. For those who deal heavily in Christian material culture, some are devout adherents of a particular tradition, others identify in the "spiritual but not religious" vein, and still others are non-theist or atheist. The stories of resellers shared here show selling the sacred to be saturated with ethical dimensions. Ultimately, I argue that the ethics of the secondhand sacred is informed by an enchantment of materiality that accumulates because of, not in spite of, the ongoing circulation through capitalist exchange.

Sacred Waste

Resellers source inventory by "picking" through the contents of secondhand venues. The materials have been divested: detached from previous ownership, most often because of death, downsizing, discard, and/or donation. Items are available on the secondhand market because they have been passed over by singular or multiple actors. The Christian material culture that circulates can be classed as sacred waste, an ambiguous form of materiality that is both enchanted, yet also a surplus or damaged remainder (Stengs 2014; Gould 2018). For certain kinds of sacred waste, such as consecrated materials in a permanently shuttered church, formal ritual protocols provide instructions for handling, preservation, or disposal (Myrvold 2010; Cuperus 2019). For other items, such as divested religious objects, no such protocol exists. The absence of authorized procedures poses practical and ethical dilemmas for how people engage with sacred waste (Stengs 2014, 237).

Like other kinds of secondhand material, sacred waste accumulates value through its circulation across space, time, and contexts of ownership (Appelgren and Bohlin 2015). Enchantment is "sticky" to begin with, as sacred origins are difficult to sever (Cuperus 2019; Stengs 2023). But there is also a cumulative effect at work; enchantment accrues through ongoing use and performance, as materials enter in and out of the ritual life of individuals and communities. I observed earlier that there is an irony surrounding sacred commodification. There is a long history of, and a series of robust industries exclusively devoted to, selling the sacred (cf. Einstein and McFarland Taylor 2024). Yet, there is also a widespread ambivalence that seems permanently adhered to this form of capitalist exchange (Stengs 2023, 7). This ambivalence has surfaced in multiple ways during my fieldwork. Nearly every reseller I have met has shared a story about being offered Bibles, prayer books, and other "religious items" for free because someone else professed a desire to either not profit from them or avoid them being placed in a garbage bin. This same impulse is mirrored by signs like Fig. 13.1, versions of which are found in both non-profit and faith-based thrift stores.

The ambivalence toward sacred commodification is also marked in brief conversations and jokes exchanged at secondhand venues. In March 2022, I visited an estate sale in the Chicagoland village of Brookfield. The modest home had belonged to a Croatian Catholic woman and there was a table of "religious décor" in the basement staged by the estate sale company. A box was filled with prayer cards and pamphlets, probably ~150 individual pieces. I asked the man running the sale how much they wanted for the entire box. He offered $10, which I considered an incredible bargain. I happily agreed and he quipped in reply, "Now, because I got you that good deal I'm going to get some blessings." This prompted a solid laugh from myself and others in earshot.

Consider also the mixed register of ir/reverence non-Christian resellers use when talking about secondhand sacred finds. Emily is a part-time

Fig. 13.1 Sign inside a St. Vincent de Paul thrift store in Aurora, Illinois (photo by author)

reseller who lives near Syracuse, New York. We met through Instagram after she posted a "religious drop" for sale in April 2022 (Fig. 13.2).

Emily's main strategy for selling on Instagram is to collect items by theme, then advertise when the items will be posted one by one. During our first interview, I learned that "religious" materials are among her favorite to source and sell. She can acquire them inexpensively, they sell quickly and reliably turn a worthwhile profit. Emily is an atheist, ex-Catholic who values the aesthetic beauty of Christian material culture, appreciates the opportunity to "rehome" sacred items with people who

Fig. 13.2 Screenshot of one of Emily's Instagram religious drops (screenshot by author)

are devout, but harbors no personal devotional commitments. This complicated relationship is evident in her humorous item descriptions. For example, during a June 2022 religious drop she posted a trinket box adorned with an image of the Last Supper and wrote, "perfect spot to stash your stash." For a rosary sealed in its original package she included this note: "I didn't open this up, so it's a bit of a mystery what it looks like! I didn't want to negate any of the blessings with my sinful hands ☺."

The accumulated enchantment of sacred waste resonates with a broader quality of material circulation. Colwell (2022) presents this as a palimpsest theory of objects, in which the movement across space and time entails processes that affect value, meaning, and use. Colwell highlights four processes: inscription (imprints of physical change); erasure (physical or discursive removal); action (physical change due to recontextualization); and, context (shifts in meaning due to recontextualization). Erasure plays a primary role in the social life of the secondhand sacred. So much of an object's biography is lost or obscured as it travels from owner to owner. Consider something like an antique Bible discovered on a thrift store shelf. Its previous chain of owners and its contexts of use are difficult, if not impossible, to know. A similar Bible found at an estate sale promises a bit more context: insight about the other kinds of materials the owner possessed and details about the owner themselves. What can be gathered is far from a robust ethnographic account, but rather a portrait that is partial and suggestive at best, and misleading at worst.

While so much is unavailable due to erasure, a series of potential inscriptions work like enticing clues. Original inscriptions, such as a maker's mark and a date or place of production offer details of object biography. A previous owner may have written their name. They may also have scribbled marginalia. A note explaining a gift reveals the stated intention for why an object was given, the social relationship the gift was part of, and the date of exchange. Inscriptions can also be less overt. A chalk ware statue might be chipped in multiple spots. The cover print of a prayer book might be faded from years of devotional handling. These evidences of use are valued by resellers because they are testimonies of age, use, and "character." In December 2022, I posted a chalk ware Mary statue to my Instagram account, a Salvation Army thrift store find in northeast Chicago with pieces missing from her fingers and cloak. One of my ethnographic collaborators commented in reply: "perfect level of endearing chippiness ☺."

Part of the reseller craft is to create an object narrative based on these suggestive inscriptions and contextual cues (cf. Douma 2015). The reseller who has taught me most about this is a 40-year-old white man named Mark. Based in a southwestern Chicagoland suburb, he began collecting and reselling five years before we met. He works as an IT consultant for a health insurance company, which pays the bills, but is not his passion. What Mark loves to do, what he spends most weekends and many weeknights doing, is scavenge secondhand venues for "treasures." Mark is an atheist, but a prominent vein in his sourcing centers on "vintage Catholic and other super cool Christian shit." He is drawn to the artistic beauty of items, to the historical and individual stories they embody, and to their resonance with "curiosities," his unifying category of interest.

In May 2022, Mark and I visited an African Methodist Episcopal church estate sale in Chicago's Edgewater neighborhood. The church building opened in 1891, and was granted inclusion on the National Register of Historic Places in 2008. By 2021 the congregation had dwindled from

600 members to less than 40 and could no longer sustain the infrastructural upkeep. This church's story represents an increasingly common phenomenon: historic urban churches, especially majority black congregations, endangered by gentrification processes—long-term residents displaced and church memberships declining to the point of unsustainability (Sadar and Maldonado-Estrada 2021; Bruce 2023).

This estate sale was of a particular sort: the kind where nearly everything inside was for sale. A few liturgical items had been set aside for relocation to other churches, but all else was up for grabs. Even if it was attached to a wall or a floor, if you could get it out you could buy it. I arrived early to take photographs of the property, and Mark met me onsite soon after the opening time. He bounded up in his usual manner: beaming, excited to begin, and hopeful that something surprising was waiting. We spent an hour exploring rooms, picking through book cases, file cabinets, and closets. Mark spent $80 on a small pile, including a plastic nativity set. His eBay listing of this set is a narrative performance of partial object biography:

A vintage blow mold set, was the display set for a methodist [sic] church in Chicago. Saved it out of the church prior to its demolition. In good shape, retains original paint, can use a bit of TLC. I'm not seeing any breaks or chips... this set is hard to find intact! Happing bidding!! ☺.

Object narratives like Mark's are designed to add monetary value through affective backstory. They also become part of the circulatory dynamics of sacred waste, at least temporarily attached to the object as it awaits a next home. Bibles, chalk ware Mary statues, plastic nativity sets, and all other secondhand sacred materials testify to the palimpsest theory of objects, as their social life is always being shaped by the intertwined processes of inscription, erasure, and recontextualization. These qualities of circulation set the stage for the ethical considerations that resellers confront. We turn now to those considerations, and the ways in which accumulating enchantment informs ethical engagement.

Secondhand Sacred Ethics

Resellers are not a community of practice with an authorized standard of ethics that people turn to for guidance. While they often raise ethical questions publicly through forums like social media, and they are consistently saturated in ethical considerations, there are no centralized institutional sources for their questions. The result is a centrifugal splintering of stances that cohere around particular formations of identity and ideology.

Negotiating Ethical Limits

Consider resellers who are more theologically conservative Christians. These individuals map their theological boundary marking onto the kinds of materials they are willing to source and sell. Ruth is a charismatic catholic in her late 70 s who has been circulating sacred items for over 20 years. She sees herself as "rescuing" materials that have been discarded, intervening in their social life before they might head toward a landfill. When touring her home in the Chicagoland village of Mount Prospect, she showed me dozens of saint statues that she found in garage sales or placed in alleyways for garbage pickup. If they are damaged, she repairs them. When I asked if she avoids any particular class of materials, she immediately said that anything "occult" is off limits. She named paper materials that discuss reincarnation or "amulets" from non-Christian religions. Jody, an Assemblies of God Pentecostal from Oregon, said much the same. She resells as a service for her local church library and refuses to take in any "demonic books or products," which for her includes things like Islamic materials, Mormon materials, and anything vampire-themed. Christian sex manuals pose something of a dilemma because it is not always clear how they are theologically situated. In this case, she consults Christianbooks.com as an authorizing source—if they do not sell it, neither will she. Aaron is a white man in his late 50s and the pastor of an evangelical church in southwestern Ohio. He started reselling in 2007 as a way to earn extra money to help pay for their daughter's wedding. When I asked Aaron if there were materials he considered off limits, he was quite clear. He regularly encounters Hindu, Mormon, and Freemason materials, but does not like supporting their circulation, even if they could be profitable. This is not about theological agreement; he will sell Catholic and Prosperity Gospel materials. His ethical orientation is a material reiteration of which traditions and doctrines he considers inside or outside of God's Kingdom.

The question of ethical avoidance figures differently among atheist resellers. Consider Emily, the part-time seller in Syracuse we met earlier. There are materials she assiduously avoids: illegally traded materials, such as ivory, or racialized imagery, such as animalistic representations of African Americans. During our first interview, she recalled buying a stack of vintage postcards on eBay to include as gifts for buyers. The stack included a confederate flag card, which she promptly ripped up. However, she abides by no such limits with Christian material culture. When I was sifting through books at an estate sale in June 2022, I stumbled on a conservative Christian paperback, *Selling Jesus*. I posted an image of the cover to Instagram with the caption, "if this is a sign, I'm ignoring it." Within minutes, Emily sent me a private message, which we followed up on in our next interview. Her response took my quip in a more serious direction:

> I have zero qualms of selling religious items. I was raised strict Catholic; a priest at my school was convicted of molesting kids. I watched nuns and priests belittle and verbally abuse my gay friends in high school. I still struggle with issues of shame and guilt even though I left the church 25 years ago. To me, selling religious items is just a small bit of payback for years of suffering.

Emily sources, sells, and profits on Catholic materials not out of devotion and not only out of a calculus of margins, but out of trauma.

Emily will also transform Christian material culture in a way that Ruth, Jody, and Aaron would not. Emily is a mixed media artist by training, and works as a public-school art teacher. Like many resellers, she encounters quite a bit of Catholic material when sourcing. One thing she pays attention to is the potential for "upcycling": materially transforming a sourced find in a way that adds or generates new value. Her first attempt was a white bisque bust of Mary, which she painted in the Mexican tradition of *Dia de los Muertos* (see Fig. 13.3).

The transformation took Emily three hours to complete. It sold immediately on Instagram for $36 plus shipping and she received multiple commissions to create similar pieces. The idea to upcycle came from two sources. First, while she has no known Mexican ancestry, the aesthetic is one of her favorites and she displays several *Dia de los Muertos* artworks in her home. Second, there is dragjesusglitter: an artist account she started following on Instagram a few years ago. This artist sources holy cards and prints with popular representations of Jesus from thrift stores and remakes them in the image of a "yaaas king," "the higher the eyebrow arch, the closer to God" (see Fig. 13.4).

Emily was inspired. She recognized herself in what dragjesusglitter was doing: using the traditional to be irreverent, creating unanticipated beauty from widely beloved images, making the mass-produced more artisanal. Upcycling is a way to exercise creative talent, but it is also an ethical

Fig. 13.3 Upcycled Mary bust sold on Instagram (screenshot by author)

decision: treating the secondhand sacred as a materiality that can be altered and recontextualized in a different ideological register.

Mark, the lover of "vintage Catholic" we met earlier, is an omnivorous seller. He sells plastic nativity sets, authenticated relics of saints, Orthodox icons, fundamentalist Protestant tracts on faith healing, and much more. If it is a promising monetary flip, he is game. If it meshes with the "oddities and curiosities" aesthetic he maintains for his inventory and his Instagram feed, he is game. When I first asked Mark if any materials were taboo for him, he could barely contain an exuberant "no!" However, he does attend to the ethical considerations of potential buyers. An example is Mark's response to my first encounter with a purported saint relic at an estate sale in October 2021.

The sale was in a northwest Chicago neighborhood and the first of many times I employed an ethnographic method with Mark, texting him images of materials from onsite to get his advice and feedback. He was

Fig. 13.4 Upcycled Warner Sallman Jesus print from the artist account dragjesusglitter on Instagram (screenshot by author)

quick to encourage me to buy the relic, a small square of cloth said to be from a cloak of Saint Joseph. In messaging after the sale, he offered an ethical caution: "By the way, the 'respectful way' to sell relics is to say that you are selling the holder and the theca, and including the relic for no charge. Christian customs frown on selling the actual relic for money." A year or so later, he followed his own advice (and exercised his storytelling skills) when listing a relic of Saint Anthony of Padua on eBay:

> A terrific estate find among a ton of early-1900's religious artifacts. Based on the source, I'm assuming this is also a [sic] early 1900's relic, likely 1910–1930, somewhere in there. See all pics for exact condition! Theca is in good shape. Small dent on the back, does not impact anything else of importance. Wax seal is excellent, string still intact and sealed. Front glass is in but loose. "Ex Cineribus" is "from the cinders (or ashes)." See all pics! Respecting tradition, you are purchasing the theca, relic itself is obviously included but free. No document included. I ship quickly, in a box, well packaged, and insured for items like this. Feel free to message with questions. Happy bidding! ☺.

Knowing Mark's attentiveness to ethical concerns, I wondered about his claim of having no taboos. A stress test came about four months into our ethnographic relationship. In December 2021, we met at an estate sale in the southern Chicagoland suburb of Burr Ridge. The deceased was a World War II veteran, a dentist, and a traditional Catholic (meaning he rejected the legitimacy of Vatican II reforms). Mark and I were browsing books in the garage when I noticed a stack of 9/11 conspiracy pamphlets: a sizeable stack of duplicates that appeared less like a collection and more like a tract to distribute. One of the estate sale staff remarked that the dentist was "a bit of a right winger." A few minutes later Mark found a hardback biography of David Duke, the former Ku Klux Klan Grand Wizard, Louisiana politician, and avowed Christian nationalist. Mark called me over and said, "this is too crazy not to buy."

We met at a local coffee shop afterward to debrief about the sale. I was eager, and a bit nervous, to ask about the Duke book. Mark was clear that he considers Duke's politics and worldview to be vile, but he also planned to list the book on eBay. Mark's stance is that people are "too scared" of ethically "dangerous" materials. He sees an unfortunate tendency to avoid or destroy materials rather than confront them. If anything, he takes umbrage with assertions that any particular class of materials should be off limits. Aaron and Emily, despite their many differences, do share the logic of not wanting to be complicit in circulating materials that reproduce ideologies they consider damaging. Mark inhabits a different ethical stance, akin to a stand-up comedian who wants no joke to be out of bounds.

Mark reasserted this omnivorous ethic a few months later. I had read a *Slate* article written by two historians, advocating against the practice of selling formerly confidential medical records on the secondhand market.[4] I was curious to hear Mark's take on the article and he replied with typical candor:

> Interesting article, but I disagree with almost every conclusion they make. Locking something away in a [sic] archive or vault does nothing to preserve legacy or further understanding, it hides/destroys/whitewashes/ignores. I see collection and displaying as honoring these people, not any negative connotations. You can't sanitize history like so many people are trying to do nowadays, you need to preserve and understand it, talk about it, have disagreements, take interest, or these fascinating snapshots into the past will be forgotten and that's a shame. The vast majority of collectors (aka, all) are in it for that same purpose. Displaying a photo of somebody who is disabled or

> disfigured is as much reverence as it is fascination. The alternative they propose is to shred it, destroy it, hide it, which is way more of a counterproductive stance than I think they realize.

Mark's reply is an individualistic argument. He elides the author's central objections (namely, protecting the patient's right to privacy), centering instead the personal motivations of sellers and collectors. He sees the authors as mistaking, or missing altogether, something he accepts as a taken-for-granted fact—sellers and collectors of "oddities and curiosities" are on the whole committed to ideals of respect and invested in historical preservation, not thoughtless gawkers unconcerned with de-humanizing representations. For him, secondhand circulation is preferable to alternative material trajectories.

Circulating the Good

Mark's response to the *Slate* article highlights an additional ethical dimension to secondhand circulation. His critique is that the article forecloses any potential for good that might result from reselling. This echoes a refrain I have heard repeatedly: a primary pleasure of reselling is being able to connect people with materials they treasure. Every reseller I have met fondly harbors a hope that they can "rehome" items with new owners who will truly appreciate them.

We glimpsed this with Ruth, who fashions her work as "rescuing" materials from a state of neglect to a state of reprised devotional use. Heather, a central Florida reseller, voiced a similar sentiment. Heather specializes in books, and her sales inventory lives with her personal collection numbering around 15,000 volumes. She relates to the collection/inventory in an interactive, affective manner. Most days she spends several hours sitting with her library, pulling copies to read, dust, and rearrange. She only sells books "that put good into the world," which means she refuses to list certain books on eBay. She named a copy of David Koresh's manifesto, *The Decoded Message of the Seven Seals of the Book of Revelation*. A fellow reseller suggested that she destroy the book because of its ties to religious violence, but Heather believes all books deserve preservation no matter their content.

John, a Catholic reseller in his mid-60s, offered another vantage on how reselling can foster the good. As an artist, he sources, collects, and sells reprints of Catholic art produced between 1850 and 1920. He typically sources individual holy cards from websites like Delcampe, Etsy, and eBay. Using acid-free, fine art paper and "archival quality inks," he creates various sized prints and other paper media like stickers. John said that he "loves the idea of putting out these little time capsules that won't fade away" and that his reselling work responds to "a famine in the land for the art of God." In short, John evangelizes with sacred imagery through secondhand circulation.

Aaron, the evangelical pastor we met earlier, discussed circulating the good in a monetary register. He nets roughly $12,000 each year from eBay sales. He and his wife earmark at least half for donations: to congregants in need, missionary organizations, faith-based humanitarian projects, and most recently ministries supporting soldiers wounded in the Ukraine–Russia war. For Aaron, reselling is a financial mechanism for making his faith actionable, for supporting institutions and individuals that align with his vision of the good.

An Apprentice's Dilemmas

Apprenticing as a reseller with people like Mark, Emily, and others has been a vital part of this project's methodology. Doing my own sourcing and selling allows me to ask much better ethnographic questions, and creates the basis for a robust, open-ended dialogue with my interlocutors. Because reselling is ethically saturated, my apprenticing necessarily entails a variety of ethical dilemmas.

One kind of ethical dilemma resellers confront relates to the practice of sourcing. Resellers disagree about whether or not they should maintain or divide the ordered integrity of found materials. Consider a thrifting moment from southwestern Ohio in September 2022. This particular store is one I visited regularly over the course of 18 months. It was a favored sourcing spot because I regularly found Christian material culture, from books and games to rosaries and wall hangings. There is a large U-shaped glass cabinet by the entrance, where the store displays items that have been monetarily valued at a higher price than the typical thrift donation. I rarely bought from the glass case, but it was always fun to examine.

On this September day, a middle-aged white man was closely inspecting a wooden box of slides atop the glass counter. As I passed by he was holding one of the slides up to the light. It was an image of Jesus. I stopped to inquire and he generously responded. The box was one of two Sunday School magic lantern collections, each with about 50 slides, dated

1895–1930. The boxes were priced at $100 apiece. He was a collector and antique dealer, and was leaning toward buying both boxes. I asked if he would separate the sets and sell the slides individually or if he would keep them intact. "If they come together, you keep 'em together," he replied, offering not just an answer but an ethical maxim. We ended up in the cashier's check out line at the same time, and the attendant applauded him for making the $200 purchase. The worker had done a search on his smartphone and found that individual slides were selling for as much as $12: "there's good value in that." After leaving, I sent Mark pictures I took of the slides and asked how he would have handled it. He was impressed with the slides, though he thought $200 was expensive. However, if he had acquired them, he would separate and sell individually, hoping for the $1000 profit over time. The two men voice divergent logics: one centered on maximizing profit, the other on the preservation of material integrity.

A set of 100 late nineteenth/early twentieth century magic lantern slides is an exceptional secondhand find. But, this dilemma of retaining or separating also appears in more commonplace items. A prime example is a prayer book or Bible filled with holy cards, the result of an intimate devotional practice that is widespread among Catholics and Orthodox Christians (Orsi 2018; Primiano 2022). I have encountered dozens of book-card assemblages at secondhand venues. How should they be handled? Do you leave the composed devotional object intact, cards remaining where they were lovingly tucked? Or, do you extract the cards to sell separately from the book? Again, resellers diverge. In December 2022, I visited two antique stores in Chicago's Rogers Park neighborhood. At the first, I asked to see a mid-twentieth century Catholic prayer book behind a glass case that was stuffed with cards. I asked the owner if I could separate them and only buy the cards, which he happily agreed to. At the second store, a slightly newer book was similarly filled with cards. I asked the owner the same question and he politely declined, echoing the "if it comes together, it stays together" logic.

A second kind of ethical dilemma centers on respecting the privacy of personal information. Recall the discussion of inscriptions earlier. Sacred waste materials, especially books and ephemera, often include writing by a previous owner: names, dates of acquisition, home addresses, notes explaining a gift, marginalia, to name a few. Some buyers desire "clean" copies; that is, books with no inscriptions whatsoever. For them, any writing devalues. For other buyers, inscriptions add value. They appreciate the personalized provenance. It fosters the sense that they are buying something with accumulated enchantment. Inscriptions are evidence of use, sacred remnants of an active social life. As Mark might say, they are buying a story, not just an object.

Inscriptions can pose ethical dilemmas when the writing shares sensitive information. I encountered this at a southwest Ohio thrift store in October 2022. This was an independent store, an evangelical non-profit whose proceeds support an orphanage. All Bibles are free, and the ample "religious books" section charges $0.25 for paperbacks and $0.50 for hardbacks. The inventory was dominated by fundamentalist and charismatic volumes. A prominent theme was End Times prophecy, commentaries on the Book of Revelation and apocalyptic interpretations of world events. It was on this theme that I came across Salem Kirban's 1968 *Guide to Survival*. Having sold other Kirban books, I added it to my purchase pile.

On closer inspection at home, I noticed that a previous owner had written extensive marginalia throughout the book, adding their own End Times interpretations to Kirban's. In one section, this reader had worked alongside one of Kirban's formulas for decoding a Social Security Number to test whether or not this government-issued identifier was the Sign of the Beast. What to do with a book that almost certainly includes a person's SSN? Curious for Mark's take, I messaged him a description. True to form, Mark interpreted this not as an ethical danger, but as a value-added selling point. Not only would he sell it, he would make clear in the eBay description that the marginalia include personal information. I disagreed, deciding not to list the book and instead retain it in my collection of primary sources.

A third ethical dilemma concerns materials that arguably are quite rare and, in turn, best belong in the care of trained archivists. The magic lantern slides were a rare find, but the object itself is not rare. This media circulated in the hundreds of thousands between the 1870s and 1920s, and continued in use for decades among traveling missionaries and Sunday school teachers (Schaefer 2017). Most secondhand sacred materials I encounter would not be considered rare in any professional sense. Most, but not all.

In April 2023, I visited an "oddities" store in southwest Ohio. The small shop was in transition. The owner was in the process of turning over his entire inventory, and was also scrambling to deal with a leaking roof—water being an unforgiving nemesis to antiques. I initially saw little of interest, until my eyes landed on a vinyl LP. The front image of a choir registered that it was some form of Christian material culture and then my brain processed the cover print, "People's Temple Choir." Still in its original plastic wrapping, the LP was from the commune led by Jim Jones. The owner assured me this was an original pressing of the album and he no doubt registered my excitement. He made an offer of $225, and later came down to $200. My sourcing for resale rarely goes above $10 per item, and sourcing for collecting goes only a little higher. I left empty-handed, eager to know more about the album.

Settled at home, I pieced together some of the story. For starters, the music is available for free through multiple online venues. Second, the music is lively, infectiously catchy, and, to my tastes, quite good. It is the only album the community every produced; released in 1973 while still in California and five years before the mass suicide in Guyana (Harvey 2021). A 2018 *Rolling Stone* profile of the album describes it as a 12-song collection of "gospel-pop-funk," and also notes that "the original vinyl version of *He's Able* is very scarce."[5] As of this writing in January 2024, one copy is available online—$500 or best offer on eBay.

While the music itself is only a Google search away, the original LP is far more limited. In hindsight, $200 might have been a better offer and the album rarer than I realized. I regret not buying it, though not for resale or as an addition to my collection of nineteenth- and twentieth-century Christian material culture. I regret missing the opportunity to acquire a sought-after piece of American religious history that could be donated to a museum or library archive where the LP would complement existing collections and where preservation was more secure. The sealed copy of *He's Able* presented an ethical dilemma in which the imperatives of resale conflicted with the imperatives of scholarly curation.

Conclusion

Christian material culture is circulating through secondhand networks at a massive, frankly immeasurable scale. Log onto eBay or Instagram, and you find an incredible range of items. The dizzying abundance continues at local thrift stores, where shelves are restocked multiple times daily, and on the estate sale circuit, where hundreds of homes are opened every week to transform personal possessions into negotiable commodities. This chapter has taken up a particular dynamic in this ongoing process: the ethical dimensions of circulating the secondhand sacred. The central observation I have sought to develop is that reselling is an ethically saturated practice, and that the accumulated enchantment that attaches to sacred waste informs the ethical considerations resellers make on a daily basis.

This chapter has featured resellers who are separated geographically and who are tied together as a community of practice through shared activities and shared participation in online platforms. This offers a useful contrast with other accounts of sacred waste. Consider Cuperus (2019), whose ethnographic work was grounded primarily in one city (Utrecht, Netherlands) and focused on the circulation of materials from one tradition (Catholicism) across three contexts: church, museum, and market. Churches and museums diverge in how they heritagize items, the latter introducing secular categories such as importance to national history and identity (Meyer 2019). But, they both "restrict [objects'] movement, often with the explicit goal to prevent items from ending up on 'the market'" (68). Ethical considerations in these two contexts are closely structured by the protocols set by authorizing institutions for how to handle, display, store, transfer, and otherwise interact with objects.

To examine the market context where sacred materials become commodities, Cuperus takes us to Relimarkt, an antique store in the far southeastern corner of the Netherlands. Here, a large accumulation of Catholic materials is for sale in a space that integrates ecclesial, museological, and capitalist elements. While the owner is non-religious, his ethical engagement with sacred waste reveals a sentiment that buyers should intend respectful uses. Relimarkt inches us closer to the U.S. resellers analyzed here, as both operate outside of formal procedures. In the absence of codified rules, there is a centrifugal splintering of ethical stances, which are in turn informed by widely shared dispositions, such as theological commitment, trauma, and individualist ideologies. These ethical stances, as much as any other economic or infrastructural factor, are key to understanding the circulation of sacred waste across time, space, and contexts of ownership.

The secondhand sacred conjures longstanding anxieties about the relation between capitalism and religion. In one view, commodification is an eroding force. In her analysis of kitsch as a treasured cultural form, Olalquiaga upholds this perspective: "Like fallen angels, objects lose or rather ruin their auras upon descent, arriving with little more than a crumbling, dusty shadow of their once iridescent haloes" (Olalquiaga 1998, 95). The analysis of reselling presented in this chapter suggests otherwise. Sacredness not only sticks to Christian material culture, it accumulates through an ongoing social life of commodified exchange. This accumulated enchantment sets the stage for ethical engagements, which ultimately play a pivotal role in the continued circulation of the secondhand sacred.

Notes

1. For example: https://www.christianpost.com/news/charles-stanleys-grandson-selling-famed-preachers-items-on-ebay.html (Accessed 8 August 2023).
2. For example: https://www.dailymail.co.uk/news/article-12055963/Charles-Stanleys-grandson-grandfather-personal-belongings-RETURNED-sold-eBay.html (Accessed 26 January 2024).
3. This chapter emerges from ongoing fieldwork in face-to-face and online contexts. The research design has several dimensions. I have conducted semi-structured interviews with 30 resellers. With several interviewees, I have toured homes and brick-and-mortar secondhand stores to observe collections and inventories. With one interviewee, we have attended estate sales together in order to observe how he sources materials in this context. Online, I trace the circulation of Christian material culture on multiple platforms and observe Instagram live sales that exclusively or heavily feature Christian material culture. Finally, I have apprenticed as a reseller, sourcing from 30 estate sales and more than 150 secondhand venues. From July 2021 to June 2023, I made 530 sales on eBay and communicated with more than 350 buyers about their purchases.
4. See: https://slate.com/technology/2022/07/vintage-asylum-records-found-on-ebay-history-of-disability.html (Accessed 26 January 2024).
5. Source: https://www.rollingstone.com/music/music-features/jonestown-jim-jones-peoples-temple-choir-gospel-album-746873/ (Accessed 26 January 2024).

References

Appelgren, S., and A. Bohlin. 2015. Growing in Motion: The Circulation of Used Things on Second-Hand Markets. *Culture Unbound* 7:143–168.

Bruce, T. C. 2023. Placemaking in a Postsecular Age: Sorting "Sacred" From "Profane" in the Adaptive Reuse of Relegated U.S. Catholic Churches. *U.S. Catholic Historian* 41 (1): 93–115.

Cantwell, C. D. 2017. "Religion...Is Our Business": Religious Workers and Religious Work at the David C. Cook Publishing Company. *Practical Matters* 10:1–16.

Colwell, C. 2022. A Palimpsest Theory of Objects. *Current Anthropology* 63 (2): 129–157.

Cuperus, J. 2019. *The Sacred Lives of Things: Valuing Church Objects as Heritage and Commodities.* MA thesis, Department of Religious Studies. Utrecht: Utrecht University.

Douma, M. J. 2015. Sorting the Past: The Social Function of Antique Stores as Centers for the Production of Local History. *International Journal of Regional and Local History* 10:1–19.

Einstein, M., and S. McFarland Taylor, eds. 2024. *Selling the Sacred: Religion and Marketing From Crossfit to QAnon*. London: Routledge.

Gould, H. 2018. Caring for Sacred Waste: The Disposal of *butsudan* (Buddhist Altars) in Contemporary Japan. *Japanese Religions* 43 (1–2): 1–24.

Harvey, S. 2021. Jim Jones, Jonestown, and the Peoples Temple of the Disciples of Christ. In *Critical Dictionary of Apocalyptic and Millenarian Movements*, ed. J. Crossley and A. Lockhart. www.cdamm.org/articles/peoples-temple. Accessed 15 January.

McDannell, C. 1995. *Material Christianity: Religion and Popular Culture in America*. Yale University Press.

Meyer, B. 2019. Recycling the Christian Past: The Heritagization of Christianity and National Identity in The Netherlands. In *Culture, Citizenship and Human Rights*, ed. R. Buikema, 64–88. Routledge.

Myrvold, K., ed. 2010. *The Death of Sacred Texts: Ritual Disposal and Renovation of Texts in World Religions*. London: Ashgate.

Olalquiaga, C. 1998. *The Artificial Kingdom: A Treasury of the Kitsch Experience*. New York: Pantheon Books.

Orsi, R. O. 2018. *History and Presence*. Cambridge: Harvard University Press.

Primiano, L. 2022. Artifacts of Belief: Holy Cards in Roman Catholic Culture. In *Vernacular Religion: Collected Essays of Leonard Norman Primiano*, ed. D. D. Moore, 57–74. New York: NYU Press.

Sadar, C., and A. Maldonado-Estrada. 2021. *Desacralized*. DigBoston. https://digboston.com/special-feature-desacralized/. Accessed 1 October.

Schaefer, S. C. 2017. Illuminating the Divine: The Magic Lantern and Religious Pedagogy in the USA, ca. 1870-1920. *Material Religion* 13 (3): 275–300.

Stengs, I. 2014. Sacred Waste. *Material Religion* 10 (2): 235–238.

Stengs, I. 2023. Branching Into Heritage: The Genesis of the Anne Frank Tree and Its Multiplications. *Journal of the Royal Anthropological Institute* 29 (2): 363–382.

Vaca, D. 2019. *Evangelicals Incorporated: Books and the Business of Religion in America*. Cambridge: Harvard University Press.

Wharton, A. J. 2006. *Selling Jerusalem: Relics, Replicas, and Theme Parks*. University of Chicago Press.

Index[1]

[1] Note: Page numbers followed by 'n' refer to notes.

P. Kuppinger (ed.), *Second-Hand Cultures and Economies of Reuse, Repair, Sharing, and Care*,
https://doi.org/10.1007/978-3-031-99875-1

N

O

P

R

S

T

V

W

Y

Z